A Century of Aviation

Worldwide Commercial and Military

[handwritten inscription: To my oldest son Dale & family Best to you! George E. Slagley 8-2-20..]

By

George E. Slagley, P.E. (Retired)

Strategic Book Group

Strategic Book Group
P..O. Box 333
Durham, CT 06422
www.StrageticBookClub.com

ISBN: 978-1-61204-093-6

DEDICATION

This book is dedicated to the Navy , who influenced me to enter Aviation. To the staff of Parks College who , with the assistance of Mr. Harrington to obtain my degree in Aircraft Maintenance Engineering. To Mr. Conrad Busse, COL Soler, the Commanding General, and the others at the Army Aviation Command who assisted me in aviation interests. And last, but not least my parents and my wife for their cooperation my Aviation endeavors. I thank the Army for giving me the opportunity to serve our Country for over 25 years as an Aviation manager/advisor. I thank my wife and family for tolerating my long hours of endeavors and the frequent travels required for my advancement and gaining information in Aviation.

CONTENTS

TITLE   PAGE

INTRODUCTION vi

PART-I Past, Present and Future ix

CHAPTER- I Early and World War II 1

CHAPTER- II Post World War II and Korea 11

CHAPTER- III Post Korea and Viet Nam 17

CHAPTER- IV Post Viet Nam 23

CHAPTER -V Afghanistan and Iraq 28

CHAPTER -VI The Future 31

PART VII -Aircraft Companies 35

PART III- Lighter-Than-Air Vehicles 199

APPENDIX A- Definitions 207

INTRODUCTION

The development of Civil and Military aviation has been phenomenal. The civil aspect dates back to the early 1900's with the advent of mail transport and small airplanes. On the military side, the SPAD airplanes were used in World War I for observation and bombs were manually dropped from the cockpit, hoping they would reach the target. Dirigibles, termed as 'balloons" or airships were used for transport and observation. Helicopters were mostly experimental until after World War II Since then helicopters have reached a record level of use by both military and civil aviation. Technological advances have been tremendous, from the basic "fly by the seat of my pants" to the highly automated computer controlled aircraft of today. Airships have become controlled and their use is arising for observation. Drones were originally used for target practice but have became remote controlled airplanes for highly sophisticated surveillance platforms. We dreamed and read about space travel for many years, and this has become reality the men landing on the moon, assembly of a space platform, and other space missions, some, including sophisticated unmanned missions. The development of aviation has been a very interesting subject through the years. This book takes it from the early 1900's to the present to give the reader an appreciation for the technology advances and accomplishments. Also the crop dusting was accomplished by daring pilots in the thirties, forties, and fifties Many were killed or injured in accidents. In the late twenties and early thirties, several pilots were trained and provided Airmail across the United States. Charles Lindberg was an avid Airmail pilot who thought of the expansion to Europe. He did a lot of planning, and found a company to manufacture the airplane to fly from New York to Paris. He finally accomplished the mission. In the passenger arena, the DC-3 was developed by Donald Douglas, and was used for many years subsequently. It was a conventional twin-engine airplane with two radial engines and was a tail wheel configuration.

Prior to World War II, several companies were organized due to the growing interest in aviation. Igor Sikorsky, an immigrant from Russia, was the pioneer in development and success of the helicopter. Dr Werner Von Braun and some other German engineers fled from Germany during World War II and was one of the pioneers of the U.S Missile and Space program. V-1 and V-2 rockets were the beginning of the Jet Age. The Germans were the first to apply jet engine technology to aviation. In fact, the Germans were the first to design a jet fighter. The V-1 and V-2 rockets played havoc on England during World War II. The major wars that affected aviation were: World War I, 1914-1918, World War II, 1940-1945; Korea, 1950-1953; Viet Nam, 1965-1973; Afghanistan, 1991-2003; Iraq and Afghanistan presently.

Since World War II, there has been a phenomenal explosion in technology and new companies being opened and expanded. We are now in worldwide perspective and the future is expected to continue worldwide as development as the numbers of airline passengers, freight and private aviation is expected to further expand. Airports continue to expand to accommodate larger and faster airplanes.

The summaries of aircraft manufacturers are some of the most prominent producers of aircraft and not a complete list of suppliers or service businesses. They are an overview and those who desire additional information on any specific aircraft. Those interested are advised to further research the subject[s] of interest. It is interesting to note that some countries have political differences, but aviation seems to be a universal interest with cooperation in space projects and aircraft manufacturing.

Airships, including balloons and blimps, have advanced greatly and interest in sport ballooning has increased also. It was discovered that military use was limited except for long range and long flight period patrols. Helicopters and drones have overtaken balloons and blimps as a patrol platform, due to the vulnerability of airships. Home built kit manufacturers and modification companies are not listed in the manufacturing section. Only those involved in the manufacture of new commercial and military aircraft are listed.

The United States and Russia are the largest producers of airplanes and helicopters. Russian manufacturers were non-existent before the change in country politics. Aircraft manufacturing information was non-existent during the USSR years, because it was kept secret with tight controls. The reader is advised that the description of airplanes is considered adequate for the average reader. This provides a basis to guide additional research.

Every country in the world manages aviation development and production differently. Countries under Communist or Socialist control the process more intensely than others. United States Aerospace companies are open in all sources and are the easiest to research. At least two publications are supplemental to this, Chopper, by Robert F. Dorr is a good reference and the Janes All The World Aircraft series is an excellent source of additional information. Also all American companies and several foreign companies have an internet web site for additional research. Absolute up to date information is not available due to government and corporate security restrictions. The U.S military maintains strict security over contractors to insure that information is not released that may jeopardize the security of the United States. Corporations restrict information due to competition.

PART I

Past, Present and Future

CHAPTER 1

Early and World War II

Man has tried to fly by some means for many years, but the Wright brothers from Dayton Ohio were the first successful at Kitty Hawk, NC. They were the first airplane contractor to the U.S. War Department. The specification was on one page and the contract was only one page. The Wright brothers organized flying schools for the Department of War, including Maxwell Field in Montgomery AL in 1910, and the early airplanes were bi-wing; in other words had two wings, one upper and the other lower. They were all single-engine until Ford built a Tri-Motor, equipped with three engines.

Major cities built airports initially for Airmail and then due to needs of passenger traffic and crop dusting operations. Eventually smaller cities built airports to accommodate the growing interest in aviation. Schools were established to train airplane mechanics and pilots. The forerunners were Embry-Riddle Spartan and Parks. Aeronautical. Engineering was established as a separate specialty. Few mechanical engineers were trained in the design of airplanes. Aviation during World War I (1914-1918) was limited to observation of the battlefield and some dropping of bombs on targets without the benefit of bomb sights. Dropping bombs on targets without the benefit of sights was very inaccurate. Innovations, for which the U.S military is famous, created other uses for airplanes. There was also some air-to-air combat with the famous SPAD and others.

The most significant bomber of that period was the Handley Page 0//100. Aircraft. Aviation Technology was only ten years old at the start of World War I, but the technology had developed at a extraordinary pace. This first heavy bomber entered service in 1916 to attack the German Zeppelin bases that caused heavy damage to London. It was powered by two Rolls Royce engines and had a speed of 79 mph. The wingspan was 100 feet and it was able to deliver 2,000 pounds of bombs with remarkable accuracy. After World War I, the Handley was converted to a civilian airliner for the first civilian airliners of Europe.

Multi-engine airplanes were developed. The famous DC-3 by Donald Douglas was developed in the early thirties. It was one of the first to provide passenger service. It was considered the most reliable and was utilized by civil aviation and the military for many years. It had two radial engines, and a conventional main landing gear with a small tail wheel.

Flying schools and airplane manufacturers were established during the thirties. Civil aviation expanded to accepting rides at Fairs, and for people that just wanted to have the experience of flying. Air Mail continually expanded. Also manufacturers began designs for larger bombers, and fighters because of the growth of Nazism and the threat of war.

World War II caused an intensive effort to design and build bombers, fighters, and transports for the Navy and Army Air Force. Some of the first few were the twin-engine B-17 and the B-24 Liberator. Later the B-29 Flying Fortress bomber with 4 engines was deployed.

Some early fighters were the F-6 and F-8. Japan had the Zero that was used to bomb Pearl Harbor and England had the Spitfire. Germany had the Messerschmitt series. The Navy produced the PBY Catalina and PBM Mariner, which were flying boats. The main British fighter was the "Spitfire."

The B-17 and B-24 each had two radial engines. The B-29 had

four radial engines. The F-6 and F-8 were single-engine. Aircraft. The Japanese Zero was single engine radial and frequently used as a suicide vehicle. The German ME 109 was a fighter with a single engine. The Boeing B-17 was built in 1937 and was the first all metal four-engine heavy bomber.

The B-17 Flying Fortress bristled with 13 0.5 machine guns, an average bomb load of 6,000 pounds and took on the best the German Luftwaffe had to offer. They flew in formations of up to 1,000 bomber daylight raids and attacked the most heavily defended targets in Europe. However, they suffered heavy losses, but they and their crews helped turn the tide of the War in Europe by destroying the Nazi war machine. The Boeing B-29 Super Fortress had a range of over 3500 miles, an operational ceiling of 31, 850 feet and a top speed of 358 mph.. It could carry a payload of 20,000 pounds of bombs, twelve .50 caliber machine guns and a 20-mm cannon. This airplane was very advanced for the time, featuring an aerodynamic fuselage, the crew compartment was pressurized and fitted with bullet-proof glass. It was used extensively in conventional bombing missions against the Japanese, best remembered for dropping atomic bombs on Hiroshima and Nagasaki and thereby ending World War II.

The British Avro Lancaster was Britain' s most heavy bomber of World War II. It could carry up to 22,000 pounds, flew at night and pounded German cities and factories. It was one of the Bomber Command's favorite aircraft. Uses included the 1943 dam-busting strike and Battle Ship "Terpitz" sinking raids. The success of the Lancaster came with a high price. Over 55,000 crewmen were lost in the course of World War II.

The Junkers Ju-88 was probably the most important German bomber of world War II. It was in front line service through the period 1939-1945. It's versatile design enabled it to be used also as a dive bomber, torpedo bomber, heavy fighter, and night fighter. It was heavier than the Heinkel 111 and Dornier 17, and was the fastest of the Nazi bomber fleet. It was armed with seven

103 machine guns and a payload of nearly 8,000 pounds, and was a formidable opponent during its tenure.

Another was the De Havilland Mosquito nicknamed the "Wooden Wonder" was the most versatile aircraft to see action in World War II, as a bomber it was also the fastest. It was almost not detectable to radar because it was constructed of wood. Because of it's speed, it could outrun any enemy fighter. It carried no defensive armament, but was armed with a payload of 2,000to 4,000 pounds, and could fly from 10 feet to 31,000 feet altitude. It could take the fight to the enemy's front door. By the end of World War II more than 40 variants had been in action. The PBY flying boat had two engines was predominately used by the Coast Guard and the Navy. The PBM was large four-engine flying boat and was also used by the Navy and Coast Guard. Several Americans trained by the Royal Canadian Air Force and flew Canadian warplanes in combat and for transportation of troops and cargo before the U.S entered World War II. Some transferred to the U. S Army Air Corps after the U.S entered the war in Europe.

A man named Norden invented a bombsight for the bombers, and it was used extensively and became a common term among crewmembers. It was used extensively for the bombers over Germany and crewmen were instructed to destroy it if they crashed or went down in enemy Territory.

Japan dealt a surprise attack on December 7, 1941 at Pearl Harbor. The United States was not prepared to deal with war. The Japanese were famous for their suicide missions on U.S. ships and carriers. Troops and equipment were not readily available, so American companies went on 24 hour days, seven days per week and 365 days per year. Workers were at a premium in all types of war materials, and many women and elderly were employed to manufacture war materials, including weapons, airplanes and ships. The United States initially had one aircraft carrier, and moved it at night to make the Japanese think there were more carriers. At this time, aircraft carriers were the only transportation for airplanes. Many Americans were captured by the Japanese and treated very harshly. The U.S staged from England during World War II for attacks on Germany. Many U.S planes were shot down by the massive German artillery. Some escaped but some crewmembers were captured as Prisoners Of War and treated harshly by the Nazis. All strategic manufacturers were on 24 hour days, 7 days per week operation and mostly women were trained to perform all operations including welding, manufacturing airplanes, ships, weapons and other specialties.

Men who were not eligible for the draft were also hired for jobs in the U.S. and Alaska. Military airplanes were ferried by many military and civilian pilots through South America to Africa, and then to England and other Military bases in North Africa. The reason that they could not be ferried northerly was the threat of German U-Boats in the North Atlantic This was the only mode of transportation for the airplanes, as the Navy had just started building aircraft carriers, and other air transports and ships were not large enough for the airplanes. The airplanes were badly needed by the Combat pilots and crews and as mentioned an increasing number of Aircraft manufacturers were working non-stop to develop and produce them. German U-Boats, or submarines as we know them now were spotted near the U.S. Atlantic Coast. All cruise ships and other cargo ships became part of the Navy and were used to carry men and cargo for the War. Several were sunk by the German U-boats but most made it to England and Europe to deliver badly needed soldiers and support cargo. The Coast Guard was absorbed by the Navy to support the war effort. All the ship traffic was under the Navy Department. As the campaign started in North Africa, and proceeded northward into Italy, combat airplanes were in very short supply, and initially were not used in the war effort at that time.

As soon as U.S. fighters and bombers were produced, they were used extensively from Air Bases in England and North Africa for combat duty over Germany. Most bombing missions were escorted by fighters. Many were lost, and the crews captured and interred as Prisoners Of War in Germany.

All aviation schools, including Embry-Riddle in Miami FL, Spartan in Tulsa, OK, and Parks in Cahokia IL, concentrated on training mechanics, pilots and other crewmen. Others were established to train pilots and/or mechanics.

The German government arrested all the Jewish people in Germany and most were used as slaves and/or executed. The United States arrested all the German sympathizers and aliens in the U.S and established concentration camps for them. The Americans treated them well and they were released after the war.

Many young boys purchased kits, built and flew model airplanes. This market flourished, and all 5 and 10- cent stores, drug stores stocked them. Some stores only were for model builders. As soon as a new airplane was produced, a model of it was available. The kits consisted of pre-cut balsa wood, tissue paper for the skin and small hardware. Model airplane cement was used on all to glue the pieces together. This became a hobby for most of the young people in grade school at that time. The model kits sold for 10 to 25 cents each and the model airplane cement cost 5 cents. Many eventually became engineers and technicians in the aeronautical field. Models were also available for ships, tanks and other war vehicles, and American flags were displayed everywhere, including on bicycles.

As design and production of transport airplanes, commercial aviation upgraded their fleets to the DC-4, DC-5 and DC-6. One of the most modern and popular was Constellation, commonly known as the "Super Connie." Of course the old DC-3 was still popular. The DC-3 was a conventional airplane with main landing gear and a small tail wheel. The DC-4, DC-5 and DC-6 were essentially the same except larger and went to four engines. The Super Connie had a tricycle landing gear and dual vertical rudders. All controls were through cables.

When Japan bombed Pearl Harbor on December 7, 1941 the U.S was crippled badly. They had only one Navy Aircraft Carrier in the area, and it changed positions frequently at night to fool the Japanese. It was a complete surprise to the U.S and it was widely thought that Japan had the edge on the U.S. The U.S. was crippled severely because the country was not expecting the consequences of a full-out war. This was the first time in history that bombers were destined to take off from a carrier, but could not land in the carrier and could not carry enough fuel for their return to the carrier.

Secretly an American Officer was assigned to plan for bombing of Tokyo. Lt. Col Jimmy Doolittle assembled a group of B-25 bomber crews and they trained taking off from a Carrier with a full load of fuel and bombs. Jimmy was the only one that knew what the mission was and the rest of the crews speculated on what they were being trained for. None of them guessed the correct purpose of the mission. They flew to Eglin AFB in Florida for training, then to the west coast where they were loaded onto a carrier. It was impossible for them to land on a carrier so the crews had to plan on landing on an airfield in China, near Chungking.

After their training, the carrier headed west and rumors were that they were preparing to bomb the islands where the Japanese were in control. The day before they were preparing to launch, Jimmy briefed the crews and told them they were headed to Tokyo and other cities and pinpointed the targets they were to bomb. He told them that they were to carry no documents and the targets had to be memorized. He also pinpointed possible landing sites in China if it was necessary. The opportunity was presented for any crew member to withdraw from the mission, but no one volunteered to withdraw. Standby crewmen were available, but none were used. The carrier Captain announced that a Japanese submarine and Zero aircraft were spotted on radar, and they must prepare to launch before planned. General Quarters were announced and this was the signal for the crews to man their aircraft and prepare to launch.. The navigators in the bombers were busy plotting their course for the targets. They would be over Tokyo during daylight hours instead of nighttime as planned. The pilots were advised that if they had a problem with the bomber on the flight deck, the Navy men would push it overboard to make room for the others taking off.

They were over Tokyo for a short time and encountered anti-aircraft fire. They dodged the fire and the Zero fighters were launched. They successfully bombed their targets, and to escape the Zeros and Japanese, flew into China. They were scattered all over the eastern China coast and some crashed and some were captured as POWs by the Japanese. Others landed successfully and escaped with the help of the Chinese people. China was mostly supportive of the crews and helped them to escape into friendly areas. Months later, the injured crews were rescued and flown to U.S bases. Their landings were complicated, as the Japanese had started invading China. Additional reading is the book and movie titled, "Thirty Seconds Over Tokyo" by Capt. Ted Lawson who flew the B-25 labeled "Ruptured Duck." and it is a good account of this mission.

Lt Col. Doolittle was promoted to Brigadier General after the mission and he personally took time to assist and keep the crews and spouses advised and assist them in any way possible. German V-1 and V-2 rockets pounded England and U.S bases in England. The Germans were the first to develop missiles, due the efforts and research by Dr Werner Von Braun and his team of engineers who would later defect to the United States. It was also found that they had the first known jet fighter almost ready for production. The bombing of Germany was a daily event for U.S crews. Allied troops advanced toward Berlin from the west and south and the Russians closed in from the east. The U.S Air Force was added to the U.S. Armed Services and was implemented from the old Army Air Corps. Thus, air superiority had advanced making the United States the world leader.

Some of the Navy planes used were TBM (Torpedo Bomber), F-6 and F-8. Naval aviation was growing. Blimps were used for patrolling for submarines along the West and East coasts of the

U.S. Helicopters were being developed by Sikorsky, Bell, Hiller and Kaman. According to aerodynamic principles, the helicopter wasn't supposed to ever fly, but it somehow did. The TBM had a so-called "greenhouse" where a gunner was placed. It carried torpedoes, and was aimed mostly at the German U-Boats and other German shipping.

The Army used L-1 Vigilante and L-2 Sentinel aircraft for observation and transport of patients. The Army Air Corps had P-38 and P-51 aircraft. The P-51 Mustang had an in-line engine. They also used the first helicopters, were the YR4B made by Sikorsky and the R-4, which looked like a flying shoebox with windows. It had a 200 Horsepower Warner engine, seven cylinder air cooled radial engine. It was developed by Igor Sikorsky about 1940, renamed as the Vought-Sikorsky 316A, and the first flight was in 1942. It had a three-blade rotor, with a diameter of 38 feet. The maximum speed was 75 mph, Cruise 60 mph, a range of 100 miles and combat range of 50 miles. The empty weight was 2000 pounds, takeoff weight was 2200 pounds and gross weight was 2525 pounds. During this time the helicopters were called several names including whirlybirds and eggbeaters.

The end of World War II was accomplished in Germany through the Allied forces and Russian closing in on Berlin. Adolph Hitler committed suicide with some of his close associates. Those that did not commit the suicide scattered, many of them escaping to Argentina In South America. Many thought that this victory was accomplished through air superiority. In Japan the air superiority started a turnaround. The most advanced B-29 long range Bomber was the icing on the cake. A select crew was selected for a special mission to Japan cities. The B-29 was completely loaded with a new bomb for the attack. The crew was not aware what the new bomb was nor the devastation that would reach. The crew was given the targets to bomb, and they flew the B-29 to Japan, and dropped the Atomic Bombs. The bombs erupted into a large mushroom cloud, and the glow from it was immense. The B-29 experienced buffeting from the blast.

The crew had never experienced these effects before. One pilot remarked, "Oh God What have we done now?"

They returned to base and then were told that they had dropped the first Atomic bomb. The devastation was so huge and there was a large area that suffered massive destruction. The Japanese emperor signaled the end of hostilities, and called the U.S for a peace meeting. The meeting was on a U.S. ship, and included representatives from Allied Countries and Japan. This was the end of hostilities with Japan. The destruction was the worst ever known and most hope this was the last time this type bomb will ever be used.

Another fighter developed during latter part of World War II was the F4U Corsair. It was a gull wing, single R-2800 engine, and was the fastest because of the large engine. The nose cannons were synchronized with the propeller so they could be fired through the propeller blades. It was a single-seat configuration. Also other fighters that were famous for their performance were the P-51 mustang and the P-38.

Lighter-than Air blimps were used by the Navy for observation only. They could be refueled at sea with more loiter time than any other helicopter or airplane.

England was severely pounded with German V-1 and V-2 rockets. The United States had bases in England from which attacks were launched against Germany.

CHAPTER II

Post World War II and Korea

After World War II, the threat changed to Communist Russia. They built a wall between east Berlin and west Berlin, dividing the city. This was called the Cold War and caused the U.S to put attention on development of new commercial and military airplanes. The veterans of World War II went to colleges to obtain college degrees and training under the G.I Bill of Rights. There was an abundance of aviation personnel because of their Military training and experience.

Germany had led the world on development of Jet power. Aircraft manufacturers in the United States used the knowledge and skills from Dr. Von Braun to design and market these Aircraft and Missiles. Igor Sikorsky concentrated on helicopters. At first, turboprop jets were introduced, in other words, propellers driven by jet power.

After World War II, the world threat was from Russia as mentioned earlier. A wall was built in Berlin commonly known as "The Berlin Wall," separating it into East Berlin and West Berlin. Residents were not allowed to commute between the two. The returning military pilots and mechanics were hired and retrained by the airlines. They were highly experienced, but some Civil training was required to adapt them to the airlines. Communism expanded into China and North Korea.

One of the first pure U.S Jet aircraft was the F-2H Banshee, produced by James S. McDonnell who organized McDonnell Aircraft in St Louis. It was a small fighter/trainer that formed the technical base for later models, the F-101 series, and follow-on models They also developed a jet transport that was not sold but used to transport company personnel. Dr. Von Braun and his crew and the Army Missile Command initiated missile Development. The Army, Navy and Air Force all went through reorganization. The Coast Guard was returned to the Treasury Department.

In 1947 the B-47 Stratojet first took to the skies and both the Air Force And Boeing were well satisfied with this design. It used swept wing technology captured from Nazi Germany and an unusual tricycle undercarriage, leading some to think it would only be a research airplane. By early 1948 it became clear to the Air Force that it surpassed all of its contemporaries with straight wings.

Famous test pilot Chuck Yeager was sent to follow a B-47 in a jet fighter. The next day the B-47 set a new cross-country speed record at an average of 609.8 mph to check it's speed. Chuck radioed to the B-47 pilot, "I can't keep up with you." Within a few years the airplane would become the primary bomber for the Strategic Air Command and there would be more than 2,000 built. Without the range and payload of it's successor, the B-52, the B-47 held the line as a nuclear deterrent bomber in the early years of the Cold War.

In 1947 Chuck Yaeger broke the sound barrier by flying above mach I for the first time in history. Mach I is the speed of sound, at sea level under standard conditions. Also in 1947 the US Army Air Corps became the U.S. Air Force, a separate armed service, the U.S. Army became close support aviation proponent and the U.S Air Force became proponent for fighters, bombers, transports and attack high-speed and high-altitude aircraft.

In June 1950 the U.S started a "police action" for South Korea. Because of the occupation forces around the world, the draft was implemented to accommodate South Korea. Every single young man reaching the age of 18 stood a good chance of being drafted into Military Service. The GI Bill Of Rights was continued to include Korean War veterans. This was the first unpopular war, and demonstrators protested the war. The returning veterans were scoffed at and the there emerged an attitude that servicemen were killers and not to be trusted. One of the main protesters was "Hanoi Jane." Most of the servicemen of this era remembered exactly who she was and detest her actions in and out of Korea.

This was different from previous wars, in that it was a Guerilla warfare type. There were no battle lines and it was difficult to determine who the enemy was.

The U.S. army initiated a Warrant Officer program for flight training for soldiers that could not qualify for Officers at Ft. Rucker, AL, formally Camp Rucker. This was very lucrative for recruits that had a love for aviation and could not become Commissioned Officers.

About 1952 the Air Force developed the B-36 Bomber. This was a huge bomber that was extensively tested and deployed. This had reciprocating engines, as well as helicopters being developed. Bell developed and produced the model 47 helicopter, widely used by both military and civilian entities. Hiller, Kaman, Vertol and Sikorsky also were developing helicopters, mostly of the smaller scale.

In the fifties, the U.S kept its Defense budget to counteract the Russian and Communist threat. The Navy organized anti-submarine squadrons. Blimp patrol squadrons were on the East and West coasts. Helicopter anti-submarine squadrons were on the East and West coasts . Training included the HTU bell, HTK Kaman, OH-6 Hughes, H-23 by Hiller, H-25 [HUP] by Piaseki; and H-19/HRS/HO4S, HO3S and HRP/H-21 by Sikorsky for anti-submarine operations. The Air Force and Army also utilized these helicopters. The Coast Guard had the PBY Catalina, and PBM in addition to rescue helicopters. Many of these were utilized in Korea. The Navy H-21[HRP], HO3S and HRS/HO4S were used in anti submarine operations with a sonar system that dropped a Sonar ball to listen for signals. The H-25 HUP was determined to be underpowered for anti-submarine operations. The weight of the Sonar gear plus the operator was too much. The engine was mounted in the rear, with twin rotors, one mounted on the fantail and the other mounted over the cockpit. It had one main landing gear and a small tail wheel. The H-21 [HRP] also had the same landing gear configuration. All the other helicopters had four wheels for landing, and equipped with floats or skis for special operations. The H25 HUP, H-21[HRP], HO3S, HO4S [H-19] all had hoists for lifting limited cargo or rescuing personnel or patients. The H-19/HO4S/HRS/S-55 Chickasaw was the most predominate helicopter. It was larger than any helicopter up to this time. It was equipped with a hoist for lifting. It could carry 10 to 12 passengers, eight hospital litters or a payload of 5000 pounds. The engine was accessible by opening two clamshell doors and an engine could be changed in about two hours. The engine was an 800 Horsepower, Wright R-1300 Radial piston type. The maximum speed was112 mph, cruise at 91 mph and a range of 360 miles. Endurance was in excess of five hours. Maximum takeoff weight was 7900 pounds at sea level and standard atmospheric conditions. The main rotor was three-bladed 53-foot diameter and the tail rotor diameter was 8 ft. 8 in. The main rotor blades could be folded for carrier operation or other storage by removing three tapered pins. The Marines, Navy, Air Force and Army used it. The HRS, which was a U.S. marine version, had magnesium skin which made it more vulnerable to salt water damage, so the Navy version had aluminum skin.

The HTU [Model 47], the HTK, OH-6 and H-23 were used for observation, training and light duty. They had opposed single flat-four or flat-six engines and were extensively used by Civil Aviation. The HTK was unusual because it had two intermeshing main rotors, and no tail directional rotor. The TL [47] had capability for rescue and transport of up to two patients. The OH-6 was determined the best crash survivable helicopter. The tail cone would separate in a crash, and the round cockpit would roll like a butterball. It was thought at this time helicopters could be used for observation, rescue, recreation and other light applications and were unsuitable for combat roles. This would change drastically over the next several years. Most models were used extensively in Korea and by Occupational Forces elsewhere. All were common flat-four, flat-six or radial reciprocating engines, with munitions or other equipment mounted for special purposes. They were also used for training, transportation, evacuating injured servicemen/people, rescue, and any other purpose imaginable. Howard Hughes built a huge flying Boat, called the "Spruce Goose." It was flown by Howard Hughes in 1947. He taxied, and flew a short distance, but it was never flown again, and never produced afterward. On the commercial side the DC-3 DC-4, DC-5, DC-6 and Super Constellation were the leaders Research was in process on Turboprop and Jet models, and Missile research continued to advance.

In the late 50s Turbo-Jet airplanes were produced. The Lockheed Electra was introduced for commercial and military use. The OV-1 was produced for the Army and Jet driven transport aircraft began appearing. Turbine engines for helicopters were being developed. The DC-9 was one of the first jet transports. On the larger scale was the Boeing 707. Douglas and Boeing were the leaders.

The OV-1 Mohawk was developed for the Army. It was for observation roles, and was expanded with cameras, Infrared detectors, Side looking Radar and advanced Inertial Navigation systems. It was labeled as the "Widow Maker" because it was unlikely for the crew to survive in a crash. It was used in Korea, and improvements were incorporated in subsequent models for use in Viet Nam. It was one of the first a turbo-prop airplanes. The propeller blades were shorter, because the tips of the blades could not withstand speeds approaching Mach I. The turbo-propeller airplanes were capable of higher rpm making the propeller vulnerable. Another interesting observation was loss of OV-1 averaged three per year, whether they were used in warfare or in peacetime.

The Navy relied on the improved P2V to provide patrol missions. Later models were equipped with JATO [Jet Assisted Takeoff] to enhance the short runway takeoff distance. The Army Aviation Center at Ft Rucker, originally called Camp Rucker, became the World's largest helicopter training Center. Besides the Warrant Officer program, they trained U.S officers for the Army and Air Force and officers from about 40 other countries. The Navy had four blimp [ZP] Squadrons for observation and patrol, again because of their high time that could be spent in the air without refueling and with the capability to refuel at sea if required. Some other sailors termed the blimps as "Poopie Bags."

Tu-95, the Russian bomber named the "Bear" was shown at the Moscow Air Show in 1955. It was a huge bomber designed to carry up to 4 nuclear bombs to the U.S. mainland from bases in Russia. It's existence led American planners to believe there was a bomber gap between the Soviet Union and the U.S. In reality, the Bear stretched Soviet technology to the limit, but it could still pack a big punch and for three decades was a major threat to the Western forces.

CHAPTER III

Post Korea and Viet Nam

This was the era when technology ballooned and the Armed Services took advantage of that. The Army developed the Airmobile concept, The Air Force expanded with larger and more sophisticated bombers, transports and fighters. The Navy expanded their fleet to include more advanced anti-submarine aircraft, ships, including aircraft carriers and transports. The Marine Corps improved their air power, including fighters, attack aircraft, transportation concepts and helicopter uses.

February 14, 1950 the 8th Air Rescue squadron at McCord Air Force Base, Washington was alerted that a B-36 bomber had been lost in the bad weather nearby. The weather was snowing and icy and low hanging clouds further complicated the rescue mission. A H-5 was dispatched for the rescue, but helicopters were fairly new to the public and military. They dropped ground rescue teams of three at a time, and after several days found nothing. Across the border in Canada, the search continued, to no avail. Continual search efforts finally found all but 5 of the B-36 crew and took them to safety, but never located the aircraft. It was later found out that five of the crew who had lost their lives, of 17 onboard had parachuted after severe icing and an engine fire. It was found that the B-36, the world's largest airplane was carrying a Mark IV Plutonium bomb weighing 11,000 pounds called Fat Man. This was similar to the one that was dropped on Nagasaki.

It was jettisoned over the Pacific Ocean and detonated at about 1000 feet altitude prior to the crew bailing out. There were four spare detonators and the suitcase" for a Mark IV device in the wreckage showing that the device on board lacked its plutonium warhead. For that reason, and it is still unknown why the B-36 was found where it was, and why the radar tracks and other information were in error. It was eventually found in the interior of the wilderness Of British Colombia, far from it was supposed to be.

There was much research and development of the "Fly by Wire" concept. All airplanes controls had been with pushrods and cables, but engineers thought that controlling them by electric means through electrical wiring would open up new possibilities and be safer. Missile development continued, mainly by the Army Missile Command and under the leadership of Dr. Werner Von Braun. International ballistic Missiles were installed in silos in the U.S and Russia. New threat missile guidance systems were being developed as were U.S. guidance systems. U.S aircraft were not equipped to jam foreign systems, so a concentrated effort was expedited to develop radar, infrared, and laser jammers. Some already had warning receivers, but not jammers. The main priority was infrared and radar jammers because these missiles appeared in Viet Nam as SA-7 and SA-9, developed by Russia. Research was also began to develop laser jammers to counter forecasted threats. The U.S also had a concentrated effort to obtain foreign missile and launchers to obtain technical characteristics. The U.S. knew that they could expect newer and improved missile systems. The U.S. was also developing newer missile systems, including laser guided, and they assumed that Russia was doing the same. Jet powered aircraft and helicopters came into being. Near the end of the Korean war, jet fighters emerged such as the McDonnell F-101, F-86 Saberjet, and the Russian MIG 15 series. Supersonic Jet Fighters such as the McDonnell F-4 were developed and tested. Performance characteristics of these new Fighters and Attack planes top speed were classified, and therefore will not be related herein. However, this and future jet fighters would be capable of supersonic flight above Mach 1, (the speed of sound) and Mach 2 (twice the speed of sound), etc. They could acquire and destroy a target without ever optically seeing it. The F-86 was superior to the Russian made MIG series.

In 1954, Sikorsky started production of the larger helicopter than the previous H-19. It was the H-34. It was also used for the Presidential fleet in Washington, D.C. It had a Wright 1820 Cyclone 1525 horsepower 9-cylinder air cooled radial engine. The rotor diameter was 56 feet, four-blade on a fully articulated rotor system. The tail rotor was 9 ft, 6 inches in diameter. The cockpit was above the engine about 9 feet off the ground. It had a three-point landing gear, whereas the predecessors had a four-point landing gear. The payload was 3600 pounds and the crew included two pilots and two crewmen. It had a maximum speed of 120 mph and a range of 225 miles. Armament consisted of a door-mounted M-60 or M-1 machine gun.

In 1960, a U-2 spy plane was shot down over Russia. Most were not aware of this project. The-U-2 was an extremely high altitude observation plane and it was one of the first to be nearly invisible to Radar. This created cold war tensions, but did not stop the U-2 missions. Future missions of this kind will probably be by unmanned blimps or aerial vehicles.

Bell Helicopter developed one of the first jet powered helicopters It was the famous UH-1 that would become the standard for the Armed Forces. The first flight was in 1961. It was powered by a Lycoming T-53L-13, 1400 horsepower turbo-shaft engine. It was equipped with skids or could be converted to skis for cold weather operation. The crew consisted of two pilots and two crewmen, and armament consisted of two door mounted M-60 7.62mm machine guns. It could carry 12 to 16 troops or four hospital litters. Maximum speed was 127 mph with a range of 318 miles. The military would purchase over 9000 units, and it would be used by approximately 38 foreign countries. Co-Production contracts were in effect with the Federal Republic of

Germany, Canada and China. T-53-L15 engines were also co-produced in Germany. The rotor was a two-blade, 48-foot diameter and a tail rotor diameter of 8ft 6in. The reduction gearing for the transmission had to keep the main rotor blade tip speed below the speed of sound(Mach 1).

Bell Helicopter developed the first twin-engine helicopter, the UH-1N, due to the desire have the twin engine redundancy and the demand from Canada for the twin engine UH-1. The Canadians purchased the first 50 UH-1N helicopters and the U.S armed forces also used it, again because of the twin-engine concept. Canada would later sign a co-production agreement with Bell to manufacture the UH-1N in Canada.

Another significant helicopter during this period was the CH-53, named The Jolly Green Giant built by Sikorsky. It was the largest helicopter in Viet Nam. The Ch-53A was first designated as a cargo hauler, then a combat rescue vehicle. It was powered by two General Electric T64GE-7 turbo-shaft engines and a 72ft 3in six blade main rotor. The tail rotor was 16 ft. in diameter. Maximum speed was 196 mph and cruise was173 mph.. It could haul 38 troops, 24 stretchers and 4 attendants, or 7000 pounds of cargo that could be loaded through rear doors and ramp. Up to three 7.62mm mini guns could be mounted, one in the door. The external cargo hook could lift up to 20,000 pounds It would be eventually modified to accommodate three engines. In September 1965, the first attack helicopter in the world made its first flight at Bell Helicopter. This was a giant step for the U.S. Army and changed their mission for the different Guerilla type of warfare. It was the AH-1G Cobra, with a crew of two, a pilot and a gunner. It was equipped with a T-53L-13 turbo-shaft engine with 1400 shaft horsepower. It had a 44ft 5in rotor and a range of 362 miles. Maximum speed was 200 mph maximum diving speed of 219 mph and cruise speed of 166 mph. It was initially armed with one 20mm M-197 three- barrel cannon mounted under the nose with a 750 round container; four BGM –71 Tube Launched optically tracked wire guided TOW air-to-ground missiles plus Zuni air-to-ground rockets mounted on the stub wings. The landing gear was skids, but could be modified for special missions. The Viet Cong feared the AH-1G, and when they heard it coming they would scatter and hide.

Piaseki Helicopter Co, started by the Piaseki family before 1950, was purchased by Boeing and renamed Boeing-Vertol. They developed the CH-47 Chinook, which was a large helicopter, larger than the previous H-21, H-25 and Ch-46. It had two T55-l-712 or 714 engines, mounted on the rear pylon. It had two inter-meshing 60-foot rotors and a rear-loading ramp. The range was 706 miles. There was a hydraulically powered winch for rescue and cargo handling. The armament consisted of provision for two machine guns, one starboard in the forward window crew door and one in the forward window on the port side. Maximum level speed was 177 mph and maximum cruise was 161 mph. Range was 706 miles. Fuselage was 51/52 feet long. Accommodations were for up to 55 troops, or 24 litters plus two attendants. A commercial hook system rated to carry a load of 26,000 pounds was also available. It had accommodations for air-to-air refueling. It had upgrades, as many other aircraft were subject to widen or expand the capability. Models ranged from Ch-47A to CH-47D.

Sikorsky developed the CH-54 flying crane for the Army, but the required maintenance was very high, so not many were produced. Some were used commercially for construction and oil field purposes. It used a pod system whereby pods could be picked up between the high landing gear. Different purpose pods were planned, but it was not widely used. It had a R-2800 engine and was a maintenance nightmare.

The different designations of Military and Civilian aircraft were common during this period. The civilian aircraft were designated by Model, as DC-3, DC-9, 727, 747 and DC-10. The preceding letter was the manufacturer, D was for Douglas, L was for Lockheed, etc. Air Force aircraft were designated by function such as F-101, F-4, B27, T-33, etc. The F was for Fighter, A was for attack, B was for bomber, T was for trainer, H was for Helicopter and C was for Cargo, etc. Navy aircraft were designated about the same but has designations as to type such as V for fixed wing, H for Helicopter with sub designations for purpose, such as O for observation as HO4S, which was Helicopter, Observation, model, 4, made by Sikorsky. Navy Squadrons were designated as R for transport, V for heavier than air, R for transport and Z was used for Lighter-than-Air vehicles. For example VR-1 was heavier than air transport, ZP-2 was for Lighter-than-Air Airship Patrol squadron 2, HS-3 was for Helicopter Anti-submarine Squadron 3, HU-3 was for Helicopter Utility squadron 3, VA-6 was Attack squadron 6 and VF-6 was Fighter Squadron six. The Experimental Squadrons were VX-1, VX-2, etc. Marine Squadrons were designated as VMF-6, Fixed wing, Marine Fighter squadron 6. The Army used mission designations to define the mission, such as L-19 for liaison, OV-1 for observation, OH-58 for Observation Helicopter, AH-1 for Attack Helicopter and UH-1 for Utility Helicopter.

The Boeing B-52 Stratofortress was developed in the United States, with a range of over 8,000 miles and a maximum speed of 650 mph. It had the capability to drop 70,000 pounds of bombs, and considered the most lethal bomber in the world. It could deliver nuclear weapons, Cruise missiles and precision bombs. Despite being built with 1950's technology, the B-52 would be active in the Cold War, Afghanistan and Iraq, and is expected to be in service until 2045.

CHAPTER IV

Post Viet Nam

The advanced countries of the world were advancing aviation at about the same pace that the U.S. The United States involvement in Viet Nam was the longest conflict in the history of the U.S. Most significant were the Russian MI-17 HIP and the MI-24 HIND. The HIP was a multi-purpose helicopter, larger than U.S. helicopters. It was later converted to an electronic warfare version. This version and the previous MI-8 were exported to an estimated 40 other countries including Angola, Cuba, Czech Republic, Hungary, India, North Korea, Nicaragua, Papua New Guinea, Peru, Poland, and Slovakia. There were versions for the Military, hospital, and commercial entities. The MI-8 and MI-17 HIP were equipped with two 1923 shaft horsepower Klimov TV3-117MT engines. For hot and high conditions they were supplied with Klimov TV3-117MT engines and tail rotor blades with wider chord. The options for the armed version included 128 57mm rockets in four packs or other weapons carriers, 12.7 mm machine gun in the nose, four Scorpion AT-2 'Swatter" anti-tank missiles ECM jammer, communications jammer and 23mm GSh 23 gun packs. The main rotor diameter was 69 ft and the tail rotor was almost 13 feet in diameter.

MI-24 Hind was considered the main threat to the U.S and their allies. It was designed as a gunship with capability for transport with two TV3-117 engines. The main rotor diameter was 56 ft. 9 1/2 in. and the tail rotor was 12 ft. 10in in diameter. The main rotor had five blades and the tall rotor had 3 blades. Maximum speed was 208 mph and cruise speed was 168 mph. Combat radius was 99 miles extendable to 179 miles with external fuel tanks. Range with auxiliary tanks as 310 miles to 620 miles. Maximum endurance was 4 hours. It could accommodate eight troops.

Meanwhile the U.S Army developed the Advanced Attack Helicopter, AH-64 Apache, by Hughes and later the McDonnell-Douglas after Hughes was purchased by them. Several problems developed during the early stages of the development and production. It was designed to counter the threat posed by the Russian Hind or Mil-24 and would replace theAH-1G. It went through models A through D. The armament consisted of air-to-air capability with mounting Stinger or Hellfire missiles on the stub wings. It had two GE 700-GE-701engines, each with 1696 to 1940 shaft horsepower, mounted on each side of the fuselage above the wings with armor protection for critical components. The engine had Infrared suppression built in. Main rotor diameter was 48 feet and the tail rotor was 9ft 2in diameter. The fuel capacity was 2442 pounds internal with four external tanks holding 5980 pounds. The AH-64 was continually updated from the AH-64A through the AH-64H. Performance and capability were expanded through these updates.

The Navy developed the P-3C Orion as a replacement for previous Patrol aircraft including the P2V series. They were also exported to 12 other countries and to U.S customs, Forestry, Commerce and NASA. Production lasted from 1959 until 1991. It was equipped with four turbo-prop Allison T56-A-14 Engines, each rated at 4910 hp and Hamilton Standard constant speed four-blade propellers. Total usable fuel was 9200 gallons. Oil capacity 29.4 gallons in four tanks. Mission range was 2382 miles and Ferry range was 5558 miles. Endurance was in excess of 17 hours on two engines and over 12 hours on four engines. Wingspan was over 99 feet, and overall length was almost 117 feet. Propeller diameter was 13 ft 6 in. Maximum takeoff weight was 135,000 pounds, maximum permissible weight was 142,000 pounds and maximum landing weight was 103, 880 pounds.

Armament consisted of stores in the bomb bay and on the wings. The weapons could include K-47 torpedos, MK-54 depth bomb MK-50 torpedo, B-57 nuclear depth charge, MK-82 560 lb bomb, MK-83 980 pound bomb, MK-36 and MK-40 destructors, LAU 68A with nineteen 2.75 rockets LAU-10A/C or with four five-inch rockets or SUU-44A with eight flares, and AIM-91 Sidewinder air-to-air missiles for self-defense. Maximum weapon load included six 2000 pound mines under the wings and 7252 internal load made up of two MK-101 depth bombs, four MK-44 torpedoes, pyrotechnic pistol with 12 signals, 87 sonobuoys, 100 MK 450 underwater sound signals, MK-3A marine markers, 42 MK-7 marine markers and two B.T buoys and two MK-5 parachute flares. Harpoon Missiles are standard on many P-3C. Patrol airspeed was 206 knots or 237 mph, cruise speed was 328 knots or 378 mph, and maximum level speed was 411 knots or 473 mph. Equipment included an Inertial Navigation system, Infrared jammers, sonar receivers, time code generator, electric computer controlled computer radar displays and two auxiliary computer stored data displays. It had two sonobuoys indicator sets [directional acoustic frequency analysis and recording], sonar tape recorder, MAD (Magnetic Anomaly Detector), forward computer-assisted camera, Electronic countermeasures system, and displays for two sonar operators. Equipment was continually updated to fit mission requirements.

The Airline industry expanded to larger wide-body jet transports to include the Boeing 747, the Lockheed L-1011 and the McDonnell Douglas DC-10. The 747 was the first and probably the most preferred wide-body airliner produced. This opened the market for worldwide use of airplanes for long flights. It was outfitted for several roles including passenger, Freight, and to transport the Space Shuttle. Maximum passengers were 266 with freight or 413 to 568 without. The modifications went from the basic 747, 747-100, -200B, -400, -400F and –400X. It was powered by four Pratt and Whitney, General Electric, or Rolls Royce Turbofan Engines ranging from 56,700 pounds of thrust to 62,000 pounds. Fuel capacity was 53,985 pounds. The wingspan was 211 feet and length was over 231 feet. Overall height was over 63 feet. Empty weight was 399,000 pounds, maximum takeoff weight was 0ver 800,000 pounds and maximum landing weight was 574,000 to 630,00 pounds. It had an upper deck above the main passenger area. Range was over 5,000 miles, speed mach .84. The 747-400 had accommodations for 568 passengers and increased gross weight grew to 930,000 pounds and range of 9206 miles. the trend for airlines and industry to continually upgrade the models for alternative purposes. This was partly the result of more expensive production costs and higher technology. The modified 747s were produced well into the 1990's and beyond. The world market and production facilities continually expanded and airports were forced to expand to accommodate the larger airplanes.

The famous C-130 started in 1953 and continued until 1994. The different models were C-130A, C-130B, C-130D, C-130E. HC130, C-130H, MC-130H, and AC-130U. KC-130. A total of 1066 were produced and 520 were sold to 49 countries. It had four 4508 SHP turbo-shaft Allison T-56-A-15 engines, four-blade constant speed Hamilton Standard, full feathering, reversible pitch props. Fuel was contained in six internal tanks with a capacity of 6820 gallons and two optional under-wing tanks with a capacity of 1400 gallons. Total capacity was 9620 gallons. Total oil capacity was 48 gallons. The C-130 J was powered by four 6000 SHP Allison AE-2100D3 turboprop engines, each with a Dowdy R391 six-bladed composite prop. The Lucas Aerospace FADEC propulsion system provided 31 percent more take-off thrust and was 18 percent more efficient. The propeller had 90 percent fewer parts and weighed 15 percent less.C-130 series had a wingspan of more than 132 feet and length ranged from 97 feet 9 inches to 112 feet, 9 inches. Propeller diameter was 13 feet 6 inches. Operating weight ranged from 76,469 pounds to 80,242 pounds. Maximum internal load was ranged from 38,900 to 42,673, again depending on mission configuration. One hundred fifteen [115] units were sold to 29 Nations.

Maximum cruising speed was 374 mph and runway requirement was 3580 to 5160 feet. Range was 1830 to 2532 miles and with maximum fuel 4891 miles. The KC-130 was for aerial refueling. The internal and external tanks could be used to increase range of the KC-130. This was and is the most versatile airplane ever built. The life of this airplane stretched from 1953 to well into the 2000's through many improvements, upgrades and modifications. It was and still is used for airline transport, freight, aerial tanker, aerial refueling, troop transport, special operations, search and rescue, gunship, electronic countermeasures, command control and communications, VIP transport and others. Flyaway cost was approximately $48 million. These modifications were a boon for modification contractors, and has been very profitable for Lockheed.

The latest bomber is the Grumman B-2 Stealth Bomber. It was brought into the world of advanced stealth technology with special composite laminate and secret paint. This allowed it to be almost invisible to radar and has been identified as UFOs more than any other aircraft. When based in the U.S., the B-2 on refueling missions can hit any part of the world. The cost is awesome at $2.2 billion but so is this airplane.

CHAPTER V

Afghanistan and Iraq

This era was the time of development of technology that was unheard of before. The troop strength of Korea and Viet Nam was not needed because of the technology advancement of missiles, drones, satellites, and target acquisition through infrared, radar, heads-up display, and others. Also the death rate of U.S. troops was much lower than any previous war. This can be credited to technological advancements.

Satellites and global positioning were used extensively to locate targets. This was a different war than the previous in that the enemy or foe was terrorist organizations and they used technology advancements for their terrorist activities worldwide. The National Aerospace and Space Activity continued to build the Space Station and provide supply missions to the Station. It as announced that new Space Shuttle would replace the present ones. Not much is known about the new shuttles, because they are part of the exploding technology in Military and civil areas. The airlines continued to grow with larger and better airplanes to accommodate the growth in air travel. The older turboprop airplanes were replaced with jet commuters from the smaller routes to the hubs like Atlanta, New York, etc. The hubs were used also for world travel with larger Jet aircraft. The first flights of the Airbus A-380 was the first step in several years for larger passenger carriers, and the first service is scheduled for 2007 by

Singapore Airlines. Lufthansa Airlines has ordered 15 of the A-380, entering service in 2009. Airbus has orders for 156 of the A-380 from 14 carriers, but not from any carriers in the U.S.

The A-380 is composed of about four million components produced in 30 countries. It is 239 ft. 6 inches long as compared to the 747-400 length of 231 ft. 10 inches. The Height to the tail section is a staggering 79 feet, 1 inch. It has two passenger decks running the length of the aircraft, which can accommodate 853 coach passengers. In a three-class passenger configuration, they can accommodate 555 as compared to the 747 three-class passenger load of 416. It holds 15 lavatories. The wingspan is 261 feet, 8 inches. It is referred to as the largest airliner in history. The price tag of the A-380 is 319 million dollars. Takeoff weight is 1.24 million pounds and has 310 miles of wiring throughout the airplane. It has lower fuel-burn per seat than the 747 and has 50 percent less noise, and uses less runway space than the 747. Seats are 18 inches wide, which is an inch wider than the 747. Seating is 3-4-3, which means that three seats are on the outside with 4 seats in the middle. These should be more comfortable than previous airliners, but still cramped enough to generate annoyed glances when the back is fully reclined.

The next jumbo transport will be the Boeing 787, which is projected to fly in 2008. It is touted as the most technologically advanced airliner in history. The passenger load is projected to be between 210 and 250. Cruising speed, as with all airliners will be 565 mph and range is projected 570 miles. Nicknamed the "Dreamliner" will use lightweight composites to cut fuel costs and help lessen passenger dehydration and fatigue.

Unmanned Aerial Vehicles [UAV], commonly called Drones, were used in Iraq and Afghanistan for surveillance and target location. There was a total of 1531 UAVs deployed, about 14,000 flights were flown per month and about 200,000 hours were logged. The current system weighs366 pounds, and has a wingspan of 14 feet and has an airspeed of about 95 mph. They can operate at 15,00 feet and transmit target locations for launch of missiles. Presently they are used for surveillance intelligence and rescue operations.

CHAPTER VI

The Future

The future of aviation throughout the world is expected to advance greatly. Drones were first developed for target practice. Since that time it has been discovered that they could be used for other missions. Now, drones are used for surveillance and gathering information. The future will improve these and they will be deployed to the battlefield for delivery of weapons and other duties presently reserved for manned aircraft. Helicopter drones are being developed and will most likely be used in the future to replace helicopters now used for combat and surveillance. Development of composites will continue to make aircraft and helicopters lighter, Stronger, and those warplanes encountering radar will be less susceptible to foes. The future of composites is expected to continually expand to other areas and is very promising.

Through use of technological advances, aircraft will continue to be larger, lighter and more powerful than previous models. Engines will continue to be larger, lighter, more economical and more efficient to operate.

Space travel will continue to grow with advances in technology. Many space aspects will continue to impact and improve other areas to make living more comfortable. Space trips will continue to further parts of our universe, and the resulting knowledge will be greatly expanded. Wartime casualties will be reduced, as in the past, even further as technology allows less manpower and more electrical and electronic means. Drones and improvements and improved weapons are to the point that the electronic battlefield will become a reality. Some time, the only personnel will be to insert and transport personnel, and battlefield equipment operators that perform attacks and defense well apart from the battle area. GPS will continue to get more precise in locating vehicles and personnel and will become even more precise. Aircraft, helicopter and other vehicle training aids will continue to be more precise and the fidelity will get much better. Terrain will be more realistic and readily changed to agree with the area that training is needed.

Airline passenger traffic will continue to grow, creating more challenges to the airlines and private operators. The skies will continue to become more crowded, presenting more challenges to air controllers, and greater risk to the private and corporate operators. The Federal Aviation Administration (FAA) must update to make use of GPS to control the increasing crowded Airspace.

Larger airliners will result in additional growth in commuter service. The large airliners will only operate from major hubs, and the commuter providers will expand to the more rural areas to support the large airliners and a growing customer base. The present world economy will require freight/cargo carriers to seek larger, faster and more efficient aircraft to fulfill the extra volume of perishable and non-perishable cargo product movement that will be generated.

The War on Terrorism will continue for several years. New techniques will be developed, including advanced weapons and aircraft, and this will be worldwide. Future drones will be more effective to reduce casualties.

Aircraft controls will continue to be more reliable and enhance safety. These systems have gone from cables and push-rods to electrical and now to fiber-optics. The next generation will be more advanced to allow controls to be connection-free through electronic transmittal. These will be more instantaneous and more responsive, safer, and will require far less maintenance. This will allow aircraft to be larger, faster and ease the burden on pilots and crewmembers. Speed will continue to increase. In the late 1950s Fighters and Attack airplanes surpassed the sound barrier (Mach 1) and in the future transport, freight and cargo will be moved at Mach 2 or Mach 3 and possibly Mach 4 speed with new developments in aircraft and engines. The average person will be unaware of these developments, as they will continue to be highly classified by the Military and Corporate entities. If these speeds are announced by the carriers, they will be conservative and the maximum Vne (Velocity never exceed) will never be accurate. Aircraft engines will continue to be lighter, larger, more efficient and require less maintenance. The horsepower per pound will increase and sometime another power system will replace the current jet type power.

There will be an increasing demand for mechanics, technicians, pilots and other support personnel for aircraft industry and military aviation personnel. The engineers that have the forward thinking will continue to be educated are in short supply and the supply will become more critical unless we can get more interested in aviation. The knowledge of all will be the most advanced in history. and require the latest technologies. Production Costs of aircraft will demand longer life supported by modification of present models, rather than purchase of newer aircraft.

The UAV, or Unmanned Aerial Vehicle, is starting to develop. The U.S. Military is in Advanced Development of this new concept. It will be interesting to watch this development as indications are that this will change the battlefield arena forever as well as all of our lives. This will further reduce casualties as other developments by the military have done in the past. They are presently used for surveillance and target locators, but future developments will expand these operations to armament launchers and further reduce casualties in hostile areas. Eventually, the only manned vehicles will be for troop and material delivery and pickup, and maybe high altitude bombers and control aircraft.

The use of military manned or unmanned Airships will reappear as technology advances, to be higher, bullet proof and above the reach of ground or airplane weapons. They will be used for surveillance and some other purposes which are undefined at this time. It is possible that the roles may expand to delivery of weapons and contain electronics to detect any activity and pinpoint targets based on GPS location. They will be able to be remotely controlled to acquire targets from 80,000 feet altitude and remain on surveillance for long periods. It can be assumed that steel and can withstand gunfire. Specifications are that they can operate at 68,000 feet altitude and may be used to counter Improvised Explosive Devices [IED], and do not show up on radar or emit heat for infrared acquisition. They can be used for homeland security and wireless communications.

PART II

Aircraft Companies

Here we will describe the leading aircraft companies who have contributed greatly to the advancements made thus far. Older companies like Chance-Vought, Consolidated Vultee and Glen L. Martin have became past history. Others have been absorbed or merged with other companies, reflecting the development of the large increase in Aviation in history.

AAC-Aerostar Aircraft Corporation [USA]

Aerostar is located at Spokane WA and has taken over the activities of Machen. It was formed to develop jet powered version of the Piper Aerostar and extensively upgrades the piston-engine Aerostar. AASI-Advanced Aerodynamics and Structures, Inc. [USA] Located at north Hollywood, CA, succeeded Aerodynamics And Structures formed by former Airline pilot Darius Sharifzadeh to develop the Jetcruiser. The factory was extended to 200,000 Square feet and a workforce of 300 in readiness for production.

ADI-Aero Designs, Inc. [USA]

They are located at San Antonio, TX and are designers and manufacturers of Pulsar range of light aircraft manufactured in kit form for homebuilding

AERO BOERO (ARGENTINA)

Aero Boero was established in 1952 by Hector Boero in Cordoba. They produce light civil utility and agricultural aircraft. Their focus has been on the Aero Boero 115 Trainer, Aero Boero 180 AG agricultural airplane with chemical pod and spray bars, the Aero Boero180 RVR for glider towing, the Aero Boero and AB180 PSA [Pre Selection aircraft] and AB 115 ambulance. All are one to three seat models with Lycoming O-235 or O-360. A co-production contract is in effect with Indaer-Peru, starting in 1989, under a license agreement and they remain a large company committed to providing advanced aviation technology in the Civil, Utility and Agricultural categories.

AEROMOT [BRAZIL]

Aeromot Industria Mecanico-Metalurgica. LTDA is located in Porto Alegre. They are a part of the Aeromot group with its parent company, Aeronaves e Motores, SA which was founded in 1967, and Aeroelectronica Industrial de Components Avionics SA, founded in 1981. Their latest project was the Aeromot AMT-100 Ximango and AM-200 Super Ximango. They are two-seat training and sporting motor gliders. They previously manufactured components, seats and structural parts for Embraer and seats for Airbus, Boeing, Fokker, McDonnell Douglas and other commercial transports. They have floor space of 33,368 square feet and employ about120 personnel.

AERO UNION [USA]

Located at Chico, CA, they were established in 1959 for aerial firefighting. Also, they manufacture tank systems for Douglas DC-3, DC-4, DC-6 and DC-7, Fairchild C-119, Grumman -2, Lockheed C-130/L-100 Hercules, Lockheed P-3 Orion and Electra. Also produced are aircraft fuel tanks, air stairs, environmental control systems, cargo pallet roller systems, retardant delivery and aerial spraying systems, dorsal fins, sole manufacture of Model 1080 Air refueling Stores developed by Beechcraft and acquired by Aero Union in 1985.

AERORIC [RUSSIA}

This is the Aeroric Research and Production Enterprise, and design light aircraft for production at the Sokol State Aircraft Building plant at Nizhny, Novgorod.

AEROPROGRESS [RUSSIA]

This organization in Moscow was organized in 1990 to design and manufacture utility, commuter, amphibian, aerobatic, agricultural, firefighting, training, and attack aircraft, WIG (wing in-ground-affect) vehicles, replicas and other airborne vehicles. It was originally named ROS Aero Progress and later ROKS-Aero Corporation. It is a member of the Business Aviation Association which includes the Moscow Aviation production, Yakovlev Skorost factory, the Myasishchev experimental plant, and aviation works in Komsomolsk-on-Amur, Smolensk, Noveosibirsk, Ulan-Ude and Luchovitsy. The divisions include: Utility Division, Amphibian and Special Aircraft Division, Trainer and Aerobatics Division, Business and Touring Division, Kitplane and Ultra-light Aircraft Division and Replica Division. They are mainly concerned with small private and business aircraft as well as ground-effect planes, Acrobatic airplanes and small passenger aircraft.

ROKS-Aero T-101 Gratch is a basic passenger/cargo single engine airplane powered by a 1011shp Mars TVD-10 turboprop driving an AV-24AN three blade, constant speed propeller with reverse pitch and full feathering. Three fuel tanks in each wing, each tank hold 52.8 gallons. Total fuel capacity is 317 gallons. The oil tank holds 7.9 gallons. It accommodates a crew of one or two and nine passengers or similar amount of cargo. It has a wingspan of 59 feet, 8 inches. Fuselage is 49 feet 4 1/4 inches long. Weight empty is 7342 pounds, payload is 3527 pounds and maximum fuel is 2095 pounds. Maximum takeoff and landing weight is12,125 pounds. Maximum level cruising is 164 knots or 190 mph, maximum cruising speed is 161 knots or 185 mph and maximum economical cruise is 127 knots or 146 mph. Service ceiling is 11,800 feet and range with maximum fuel is 820 miles.

T-101V is a floatplane version of the basic T-101. Wingspan is 60 feet 8 and 1/2 inches. Weight empty is 8157 pounds, maximum payload is 3,307 pounds, and maximum fuel is 2028 pounds. The floats are mounted with struts mounted to the fuselage There is a nose wheel and main wheel on each float, retracting rearward into float. Each float has a water rudder, towing point and tie-down fitting. It accommodates a crew of two and 12 passengers. Other parameters are similar as the T-101.

ROKS Aero T-106 is a twin engine development of the T-101. It has two Mars [omsk] TVD-10B or Pratt and Whitney PT6A turboprop engines, driving AV-17 three blade constant speed propellers with reverse pitch. The intended use is passenger/freight transport, ambulance, off shore patrol of 200 mile economic zones, aerial photography, Geological survey, agricultural and other general missions.

T-130 is a twin engine multipurpose amphibian with two 751hp Walter M 601E or Pratt and Whitney PT6A turboprop engines, driving V510 five-blade variable pitch propellers mounted on a single pod. Wingspan is 41 feet 3 inches, length overall is 33 feet 1 3/4 inches. Weight empty is 2755 pounds, and maximum fuel is 661 pounds. Maximum cruising level speed is 135 knots or 141 mph, and economical cruise is 105 knots or 121 mph.. Service ceiling is 16,400 feet; and range with 30 minutes reserve and maximum payload is 372 miles, and at 3280 feet altitude and maximum fuel is 820 miles.

T-175 is a single engine turboprop powered general utility aircraft. It is designed for military and civil passenger, cargo, paratroop and cargo dropping, ambulance, search and rescue, forest surveillance and firefighting, agricultural and aerial photography missions. It has a wingspan of 65 feet 3 ½ inches and the length is 49 feet 6 ½ inches. It has an operating weight empty of 7495 pounds. Maximum fuel is 2425 pounds and maximum takeoff and landing weight is 12,566 pounds and (Vne) Velocity Never Exceed is 210 knots or 242 mph. Maximum cruising speed at 3280 feet altitude is 154 knots or 177 mph. Economical cruise is 105 knots or 121 mph. Service ceiling is 13,125 feet. Range with 45 minutes reserve at 9850 feet with maximum payload is 112 miles. With maximum fuel range it is 850 miles.

T-203 PCHEL parameters are about the same as the T-201 AIST. It was designed for heavy-duty agricultural missions.

T-274 Titan is a four engine light STOL transport. It has a high wing and swept back on outer panels. Fuselage is circular with upward sweep on rear section. It has retractable landing gear, hydraulically operated, has a nose gear and main landing gear retracting inward so that they lie horizontally in the bottom of large fairings, outside of fuselage pressure cell. It is powered by four Klimov TV7-117 turboprops, each driving a six-blade propeller with spinner. It has integral fuel tanks in the outer area of each wing. There are provisions for an APU in the starboard landing gear fairing. It can accommodate a crew of two on the flight deck and payload of 5551 pounds which can be loaded with a mobile winch. It also has a hydraulically actuated rear ramp. All accommodations are pressurized and air-conditioned. The wingspan is 104 feet 7 ½ inches, length is 92 feet 1   inches and rear loading door length is 23 feet 3 ½ inches. Width is 7 feet 10 ½ inches. It has a maximum payload of 28,660 pounds, maximum ramp weight is 80,465 pounds and maximum takeoff weight is 79,365 pounds. The maximum speed at 19,685 feet is 340 knots or 391 mph; nominal cruising speed at 26,250 feet altitude is 323 knots or 372 mph; service ceiling is 30,500 feet and range with 60 minutes fuel reserve and maximum payload is 560 miles and with maximum fuel, the range is 4100 miles.

T-401 SOKOL is a single engine light multi-purpose aircraft. It is designed for passenger and freight transport, ambulance, offshore patrol, primary training, and agricultural missions. It is a high wing monoplane. Power is obtained with a 355hp nine-cylinder air-cooled radial engine or a Textron Lycoming piston engine. The propeller is a three blade constant speed configuration. Fuel capacity is 92.5 gallons. wingspan is 44 feet 9   inches, and overall length is 29 feet 2 ½ inches. It has an all metal, square section, semi-mono-coque fuselage and a conventional two-spar wing. The landing gear is non-retractable tricycle, with fairing over each wheel. It can accommodate six persons including one or two pilots. All passenger seats are removable for conversion for freight. The ambulance option carries one stretcher patient, one or two medical attendants and medical equipment including an 8 narcosis apparatus, Ki-4 oxygen supply, or a Volga automatic respirator, breathing apparatus, electro-cardio stimulator. There are also provisions for photographic and video equipment, as well as a PZS-68 public address system and various work stations. Empty weight is 3153 pounds, maximum payload is 992 pounds, maximum fuel is 705 Pounds and maximum takeoff and landing weight is 4475 pounds. Maximum level speed at 9840 feet altitude is 156 knots or 180mph, maximum cruising speed at 9840 feet altitude is 145 knots or 167mph and economical cruising at 9840 feet altitude is 116 knots or 134 mph. Service ceiling is 13,125 feet and range with maximum payload and 30 minute reserves is 485 miles, and with maximum fuel it is 1012 miles. Takeoff distance is 4100 feet and landing distance is 3775 feet.

ROKS-Aero T-407 is a single engine light multipurpose aircraft with conventional design strut braced high wing monoplane of extremely simple design. It is all-metal with steel tube engine mounting and rear fuselage truss structure. The landing gear is non-retractable tricycle type with a single wheel on each unit. It is powered with one 355hp M-14P nine cylinder radial engine driving a V530TA-D35 three blade propeller. Fuel tanks in outer wings old 83.5 gallons. It has basic seating for one or two persons on the flight deck with up to five in the cabin on two rearward facing seats and a rear bench seat. Baggage space is behind rear seats. There are alternative configurations for freight, ambulance, agricultural, off shore patrol and other utility missions. Wing span is 39 feet 9 inches, length overall is 32 feet 9 inches and weight empty is 2775 pounds. Maximum fuel is 661 pounds, with maximum payload 331 pounds, and payload with maximum fuel is 992 pounds. Maximum takeoff weight is 4585 pounds and maximum cruising speed is 116 knots or 133 mph. Nominal cruising height is 6560 feet, takeoff run is 1575 feet, landing run is 952 feet, and range with 45 minutes reserve and maximum payload is 472 miles and with maximum fuel 1105 miles.

ROKS-AERO T-433 Flamingo is a single engine light multi purpose amphibian. It may be used for passenger/cargo, civil variants for economic zone patrol and fish survey, search and rescue, ecology monitoring and training. Military applications may include patrol and reconnaissance. It is similar to other amphibians. It has a boat hull, wide retractable landing gear, stabilizing float under each outer wing, pod on pylon above center fuselage, and non-pressurized cabin. It is powered by one 355 hp VMKB M-14P nine cylinder air-cooled piston engine, driving a V-530 two-blade constant speed propeller. The engine and propeller are mounted in the center of the aft pylon atop of the fuselage and aft of the flight deck. The wingspan is 46 feet 7 inches and overall length is 34 feet 10   inches. All other characteristics are similar to other amphibians   produced in Russia.

ROKS T-501 is a tandem two-seat turboprop basic trainer. It is a conventional low wing monoplane. It has a tricycle landing gear and is powered by one 1010hp Mars [Omsk] TVD-10B turboprop engine. Provisions under each wing accommodate drop tanks and under wing armament. It has a swept back tail and empty weight is 3240 pounds. Maximum payload is 815 pounds, maximum fuel is 485 pounds and takeoff and landing weight is 4520 pounds. Maximum level speed is 285-307 knots or 330-354 mph.. Takeoff speed is 73 knots or 84 mph and landing speed is 65 knots or 75 mph. Landing run is 625 feet and takeoff run is 525 feet. Maximum range with external fuel at high altitude is 1115 miles and at low altitude is 620 miles.

ROKS-Aero T 602 OREL [EAGLE] is a twin- engine light business airplane. It has two 355hp VMKB M-14 air-cooled radial engines, each driving a three-blade propeller. It has a wingspan of 44 feet 9   inches. Maximum fuel weight is 1585 pounds, maximum payload is 1764 pounds and maximum takeoff weight is 7055 pounds. Maximum level speed is 188 knots or 217 mph.

Maximum cruise is 172 knots or 217 mph. Takeoff run is 1247 feet and landing run is 1560 feet. Endurance with maximum payload is 7 hours and with maximum fuel 10 hours. It is a low wing monoplane with retractable tricycle landing gear, a single wheel on each main landing gear and nose gear with two wheels that retract rearward.

ROKS-Aero T-610 Voyage is a single turboprop, civil and military and multi- purpose aircraft. It can accommodate two pilots and eleven passengers. It is STOL (Short Takeoff and Landing) capable and empty weight is 4310 pounds. It has one 710 hp Mars [Omusk] TV-0-100 or one 818shp Saturn AL-34 or one Pratt and Whitney Canada PT6A-114 with a five-blade constant speed propeller with reversible thrust. It has a wingspan of 53 feet and overall length is 39 feet 7 ½ inches. Maximum takeoff weight is 8488 pounds and maximum landing weight is 7805 pounds. Takeoff run is 912 feet and landing run with reverse thrust is 840 feet and without reverse thrust is 2710 feet. Maximum range with maximum fuel and 30- minute reserve is 1150 miles. With no reserve, the maximum range is 1440 miles at 9840 feet altitude, and 1864 miles at 19,685 feet altitude.

Aeroprogress T-910 Kuryer is a twin turbofan seven-passenger business transport. It is a conventional mid wing monoplane with engines on pods of the rear fuselage and wings near the center of the fuselage. The tail is swept back, the flight controls are mechanically actuated aileron, elevators, and rudder. The tail plane has an electro-hydraulic spoiler and one piece railing edge flaps. Construction is of aluminum alloy with two spar wings and a pressurized semi-mono-coque fuselage. The landing gear is hydraulically actuated tricycle type, with a single wheel on each unit. It has electronic brakes and skid protection. Power is supplied by two turbofan engines, each generating 3750 pounds st. Fuel is contained in wings and aft of the cabin with a combined capacity of 1202 gallons. It accommodates a crew of two side-by-side on the flight deck, and 6 to 10 passengers. A toilet is in front, baggage compartment is in the rear part of the fuselage. It is equipped with air conditioning and pressurization by engine bleed air. The electrical supply is 115 volt 400 hertz, oxygen system is standard and it is equipped with an PU. Wingspan is 55 feet 9 inches, overall length is 50 feet and diameter of fuselage is 6 feet 1inch. The cabin is 17 feet 4 inches long, and 5 feet 3 inches wide and high. Empty weight is 12,566 pounds, maximum fuel is 7715 pounds, maximum takeoff weight is 21,385 pounds and. maximum landing weight is 18,30 pounds. Maximum level speed at 26,250 feet altitude is 420 knots or 484 mph, maximum cruising speed at 39,370 feet altitude is 399 knots or 460 mph, economical cruising speed at 39,370 feet altitude is 388 knots or 447 mph and stall speed with flaps down is 124 knots or 143 mph.. The takeoff run is 1640 feet and the landing run is 1313 feet. Takeoff and landing from 50 feet is 2625 feet. Range at 39,370 feet, one-hour reserve fuel and maximum payload is 2050 nautical miles or 2360 miles, and with maximum fuel, 2428 nautical miles or 2796 miles.

AERORIC [RUSSIA]

Aeroric is a designer for light aircraft for production at the Sokol aircraft Building Plant at Nizhmy, Novgorod. Their latest is the Aeroric Dingo, which is a light multipurpose aircraft with air cushioned landing gear. It is powered by one Pratt and Whitney of Canada PT6A-65B 1100shp turboprop and a Hartzell three blade B5MP-3/M10876B feathering and controllable pitch is contained in two main wing tanks, each holding 79.25 gallons; two wing tanks each holding 52.85 gallons, and one feeder tank in center section holding 39.5 gallons. Total fuel capacity is 303.7 gallons and gravity feeding is on the right side of the fuselage. It accommodates one or two pilots and seven or eight passengers. Passenger cabin is ventilated and heated. Wingspan is 46 feet 9 inches and the fuselage is 22 feet 9 inches long. Propeller is 9 feet 3 inches in diameter. Empty weight is 4943 pounds, maximum payload is 1640 pounds and maximum takeoff and landing is

8157 pounds. Maximum level speed at 6560 feet altitude is 167 knots or 192 mph. Service ceiling is 11,500 feet. Takeoff from water is 1475 feet and from land 1150 feet. Landing run is 870 feet on land and 722 feet on water. Range is 497 miles with maximum payload and 930 miles with maximum fuel.

AEROSPATIALE [FRANCE]

They are located in Paris, France, and organized in 1970 by French Government directed merger of Sud-Aviation, Nord-Aviation and SEREB. Workforce is over 45,000 including 14,000 in the Aircraft Group. They own 37 percent of Airbus Industries. The main plant is at Aerospatiale base at corporation made up of several European companies. Subsidiaries and affiliates include Aerospatiale of Canada, Aerospatiale Inc. of the United States, Aerospatiale UK, Ltd., ATR Training Center, Euro Helicopter [with DASA, Germany, Maroc Aviation [part holding], Samaero [Singapore], Societe de Constructiond' avionics de Tourisme et d'affaires [SOCATA], SOGERMA-SOCEA, Societe d'exploitation et de Constructions Aeronautiques [SECA], and Sextant Avionique. Programs include: Aerospatiale Light Aircraft (SOCATA) which is responsible for light aircraft manufacturing; Aerospatiale/ Alena ATR Series, formed in 1982 to manufacture 42 and 72 passenger turbo-prop regional transports; Aerospatiale Supersonic Aircraft, which was responsible for the SST Concord passenger Liner operated by Air France crossing the Atlantic Ocean in less than four hours; UHCA/VLCT/NLA for the studies of feasibility of a high capacity airliner seating between 500 and 800 passengers; Eurocopter, formed by the merger of Aerospatiale and MBB helicopter divisions; Euroflag which is one of the companies cooperating in development of a future military tactical transport; and EUROFLAR for development of the Eurofar tilt-rotor transport.

AGUSTA [ITALY]

Giovanni Agusta originated Agusta in 1907. They acquired license from Bell Helicopter to manufacture the Bell 47 series helicopter in 1952 and Bell has been a major partner with Agusta ever since. Their major designs are Agusta A109, A129 and A139.

A109C is a twin-engine light transport helicopter. Versions include A 109CM which is a military version of the civil 109C. It is powered by two 450 hp engines, with sliding doors and fixed landing gear, the ventral fin is removed, 109H is an 18-scout version and 109HA is 28 anti- tank version. 109EOA is an Italian army scout version. A total of about 550 of all versions have been delivered. The rotor is fully articulated with four blades made of metal. The tail rotor has two blades, and blade folding and rotor brake are optional. The controls are fully powered hydraulic, and IFR system with autopilot is available. It has a retractable tricycle landing gear. Nose wheels retract forward and caster + or - 45 degrees. Main wheels retract upward into the fuselage and the main gear has magnetic disc brakes. Emergency pop-out flotation gear and fixed snow skis are optional. The engines are Allison turbo-shafts, each rated at 450shp for 5 minutes for takeoff and 380shp for twin-engine operation. The engines are mounted side by side on the upper fuselage and separated from the passenger cabin and from each other by firewalls. Two-bladder fuel tanks in the lower rear fuselage have a combined capacity of 148 gallons, of which 145, gallons are usable. The provisions include an auxiliary tank containing 39.6 gallons of fuel. Equipment includes a 30 million candlepower searchlight and a civil sling load system. The medevac version has large upward opening bulged doors and fairings give 140 cubic feet cabin volume allowing for two stretchers across the main cabin and three seated attendants or patients. Main rotor diameter is 36 feet 1 inch and tail rotor diameter is 6 feet 7 inches. The fuselage length is 37 feet 6 inches and with rotors turning, 42 feet 9 inches. The cabin is 5 feet 4 inches long, 4 feet 9 inches wide and 4 feet 3 inches high. Baggage compartment volume is 18.4 cubic feet. Empty weight is 3503 pounds, maximum internal sling load is 2,000 pounds, maximum baggage is 331 pounds and maximum certified takeoff weight is 5997 pounds. Never exceed speed for A and B is 168 knots or 193mph. Maximum cruise is 154 knots or 177 mph for A and 152 knots or 175mph for the B. Maximum rate of climb for the A is 354 feet per minute and 315 feet per minute for B. Service ceiling is 15,000 feet for both and range is between 360 and 420 nautical miles or between 414 and 483 miles. Endurance for A is 4 hours and 20 minute sand for the B is 3 hours 50 minutes.

Agusta A109K is a military and special civil utility helicopter. 109KN is a military shipboard version, primarily for anti-ship, over the horizon surveillance and targeting, and vertical replenishment. A109K2 is a special rescue version. It is equipped with a searchlight, 440 pound winch, GPS, moving ship display and single pilot IFR system. It is made from composite rotor blades, tail rotor has stainless steel skins, optional rotor brakes, lengthened cabin to allow two stretchers fore and aft, a modified fuel system and a smaller instrument panel. The landing gear is non-retractable tricycle providing increased clearance between the fuselage and the ground. Each main strut is fixed with a V support frame. Power is provided by two turbo-shaft engines, each capable of supplying 737shp for takeoff and 632shp maximum continuous power. An engine particle separator is optional. Main rotor RPM is 384 and the tail rotor rpm is 2085. Useable fuel capacity is 198 gallons, with option for additional 39.6 gallons auxiliary tank. An optional ferry tank may be installed in the cabin. The options provide self-sealing fuel tanks, and optional closed circuit refueling system with 53-gallon capacity. Each engine has its own fuel and oil system. Optional items are rescue hoist, searchlight, cargo hook and EMS interior. Armament consists of four stores attachments, two on each side of the cabin on outriggers. Typical loads are two 7.62mm or 12.7mm gun pods, 70mm or 80mm rocket launchers, or up to eight TOW anti-armor missiles with roof mount sight, Stinger air-to-air missiles,

UAVs, plus 7.62 or 12.7mm side-firing guns in the cabin. Main rotor diameter is 36 feet 1 inch, tail rotor is 6 feet 7 inches and length of fuselage is 37 feet 6 inches. Empty weight is 3638 pounds, maximum sling load is 2,000 pounds, and maximum takeoff weight is 6283 pounds for KM/KN and 5997 for K2. Never exceed speed is 152 knots or 175 mph, maximum cruise at sea level for KM/KN is 142 knots or 163 mph and for K2 is 144 knots or 166 mph. Service ceiling for KM/KN and K2 is 20,00 feet and for OEI, KM/KN and K2 is 10,000 feet. Maximum range for KM/KN and K2 is 442 nautical miles or 509 miles, with maximum payload it is 293 nautical miles or 337 miles, and maximum endurance is 4 hours and 16 minutes.

Agusta A129 Mongoose is a light tank and scout helicopter. About 45 percent of fuselage weight and 16.1 percent of all weight is made of composites. It has a fully articulated, four-blade main rotor with blades retained by a single elastomeric bearing and retained by hydraulic drag damper and mechanical droop stop. The main transmission has an independent oil cooling system, and intermediate and tail rotor gear boxes are grease lubricated. All are designed for at least 30 minutes run dry. Flight controls are fully electronic with full manual reversion. It provides automatic heading hold, auto-hover and auto-stabilizer modes, all selected by the pilot. The gunner in the front seat has a cyclic side arm controller and normal collective lever and pedals and full access to AFCS. Power is derived from two Rolls Royce 1004 turbo-shaft engines producing 825shp each for normal twin-engine operation. They also provide intermediate contingency rating of 881 shp for one hour, maximum contingency of 944shp for 2 ½ minutes, and emergency rating of 1918shp for 20 seconds. Power input into transmission is 2,000 rpm from the RR 1004 and 23,000 rpm from the T800 engine. It has two separate fuel systems with cross-feed capability, interchangeable self-sealing and crash resistant tanks, self-sealing lines and digital fuel feed control. The tanks can be filled with foam for fire protection. There are provisions for self-ferry fuel tanks on the inboard under wing stations. Infrared suppression is standard and low noise levels are also standard. Also, up to date armament and countermeasures are installed as the mission dictates. Main rotor diameter is 39 feet, tail rotor diameter is 7 feet 7 inches, wingspan is 10 feet 6 inches, width over TOW pods is 11 feet 10 inches, overall length with rotors turning is 46 feet 11 inches and fuselage length is 40 feet 3 inches. Empty weight is 5275 pounds, maximum internal fuel load is 1653 pounds, maximum internal weapons load is 2645 pounds and maximum takeoff weight is 9039 pounds. At mission weight and at 6560 feet altitude, the dash speed is 159 knots or 183mph and maximum level speed at sea level is 135 knots or 155mph. It has a basic mission of 2 hours and 30 minutes with eight TOW missiles and 20-minute fuel reserve. Maximum endurance with no reserves is 3 hours and 5 minutes. G limits are +3.5/-0.5.

A139 was initiated by Agusta and in 1998 Bell and Agusta agreed to establish a joint venture to develop of two new aircraft, BA609 and BA139. The final agreement was signed in November 1998. Bell is the major shareholder and will undertake final assembly for AB139s delivered to North America. Agusta, which has built Bell helicopters since 1952 and is investing and participating in development of the BA609, manufacturing some components and assembling those sold in Europe and certain other parts of the world. Agusta is responsible for AB139 development and certification with participation from Bell. A military version was revealed in July 2000. Flight testing of the AB139 began in February 2001 followed by the BA609 in March 2003. The agreement provides a 75/25 percent work-share basis. The military version planned is to include armored crew seats, electronic warfare protection, infrared suppressors, two internal pintle-mounted machine guns and easily removable stub-wing weapons supports for gun pods, rocket launchers, and AAM.

Design goals include high maneuverability and agility, low pilot workload, night/all weather operation, low acoustic and infrared emissions and mission flexibility for commercial and military operators. Missions planned are offshore support, medevac, corporate/VIP transport, Search and rescue, and military operations. The main rotor is five-blade, fully articulated and ballistic tolerant. The tail rotor has four blades, flying controls are four axis, digital AFCS. The landing gear is heavy-duty retractable tricycle with twin wheels on the nose unit, and main gear has single wheels and retract into side sponsons. Power is supplied by two Pratt & Whitney of Canada turbo-shafts, each producing 1678shp for takeoff and 1530 maximum continuous. OEI ratings are 1723shp for two minutes and 1678shp continuous. Fuel tanks are behind main cabin and the transmission can run for 30 minutes without oil. It accommodates up to 25 passengers on crashworthy seats in three rows of five, two forward facing, one rearward facing in an unobstructed cabin with a flat floor. A flight accessible baggage compartment is at the rear of the cabin alternatively, six stretchers and four attendants are in the Medevac configuration. Design also includes systems duplicated and separated, main and tail rotor ice protection is optional, HUMS, and chip detection for main, intermediate, and tail rotor gearboxes. Main rotor diameter is 45 feet, fuselage length is 41 feet, width is 10 feet 6 inches, and height is 6 feet 2 inches. Takeoff weight is 16,086 pounds, internal payload is 6702 pounds and external payload is 7239 pounds. Maximum cruising speed is 157 knots or 180mph, economic cruising speed is 150 knots or 172mph and endurance is 3 hours and 54 minutes.

AIRBUS [FRANCE]

Airbus was organized in 1970 to manage development, manufacture, marketing and support of A-300 series, A-310, A-319, A-320, A-321, A-330, A-340 and A-3XX. Airbus is located at Celex, France. Airframe prime contractors were Aerospatiale, which owns 37.9 percent interest, Deutshe Aerospace has 37.9 percent, British Airways 20 percent, and CASA 4.2 percent. Most notable is the A-380, which is the largest passenger plane ever built. Subsidiaries include Ae formation and Airbus industry of North America, Airbus Industrie China of Beijing in 1994 with training centers and spares store. Xian Aircraft Company and Shenang Aircraft Corporation make parts for Airbus aircraft. In 1994 Airbus signed an agreement with IPTN to assist in flight testing their turboprop aircraft [Indonesia] between 1994 and certification in 1997.

Airbus joined its four member companies in 1993 to study a Very Large Commercial Transport with Boeing. This evolved to be the A-380, which is the largest airliner. Airplane deliveries were 163 in 1991 to 157 in 1992 and 138 in 1993. The first A-340 was delivered to Air France. This was Air France's 1000th Airbus airliner. The backlog was 67 at the end of 1993. Planned output was 138 in 1995 and 154 in 1996. The A-300 was used throughout the World as an airliner. The billion in 2006, and is expected to double in the next 15 years. Airbus is expected to grow with the increase in passenger traffic. Their employment is over 2000, with about 35,000 directly employed by its five partners. Each A-380 consists of about 4 million components produced from 30 countries. Cost per unit is now 319 Million dollars and it has 50 percent less noise than the 747-400. Of course Airbus will continue their programs with smaller airliners.

AIR & SPACE (USA)

Located at Paducah, KY, this is a Manufacturing and marketing subsidiary of Farrington Aircraft Corporation. The Air & Space 18A is a two-seat autogyro.

AIR TRACTOR [USA]

This company is located at Olney, TX and produces agricultural aircraft. The design is based on 35 years of experience of Leland Snow who produced Snow S-2 series which later became Rockwell S-2R. They build several models, powered by Pratt & Whitney

PT-6A and R-1340 engines. They are one or two seat and are used for crop spraying, and fire fighting.

AMERICAN CHAMPION [USA]

American Champion is located at Rochester, Wisconsin. They offer one and two seat Explorer, Super Decathlon, Scout, formerly Marketed by Bellanca Aircraft Corporation and then Champion Aircraft Company. In addition, they provide new metal; spar wing for retrofit existing 8GCBC aircraft. Production averages 35-40 aircraft per year. The engines are used are Lycoming O-320-series or AEIO-360 series flat four engine, AEIO360. The two-seat aircraft are for training, touring, aerobatics, and utility.

AMERICAN EUROCOPTER [USA]

This company located at Grand Prairie TX combined the former MBB helicopter of West Chester PA and Aerospatiale Helicopters of Grand Prairie TX, both which had modification and production as well as sales. MBB plant is now concentrating on support. They proposed a two engine AS-350BA for Army training aircraft, which was awarded to Bell as the TH-67 Creek.

AMERICAN GENERAL [USA]

American General had production and marketing rights for Gulfstream Aerospace AA-1 lynx, AA-5B Tiger and GA-7 Cougar, bought from Gulfstream in June 1989. Trainer version of the Tiger produced as AG-5B. Agreement 1991 for license manufacture of GA-7 Couger by Tbilisi Aircraft manufacturing Association in former Soviet Dimitriov aircraft factory. In 1994, American General ceased operation and vacated the Greenville, Mississippi plant. They were in discussion with investors about restarting another Tiger production location. The Tiger is a four-seat private aircraft with a Lycoming O-360 engine. The principle market was the Florida Institute of Technology School of Aeronautics.

AMF [UK]

AMF Aviation Enterprises, LTD is located at Membury Airfield, Lambourne Berkshire. They were organized to manufacture the Chevron range of light aircraft. The AMF Chevron 2-32C is a two seat\ side-by-side micro light trainer. It is certificated in Germany, Japan and U.S. The AMF Chevron 2-45CS is another variant which is smaller and has a higher rate of climb and the roll rate has been improved.

ANTONOV [UKRAINE]

Antonov OKB was formed in 1946 by Oleg Konstantinovich. He died in 1984 at age 78. Latest productions are An-32 at Kiev, An-72/74 at Kharkov, An-124 at Ulyanovsk. Small An-2 and An-28 is built by PZL Mielec, Poland plus small batch of AN-2 [as Y-5B] in China. More than 1600 have been exported to more than 42 countries. NATO names are: Colt, Cub, Coke, Curl, Cash Cline, Coaler, Condor and Cossak. Types produced are transport, Sigint, patrol, and Airliner. The largest one is the An-225, which can carry the Space Shuttle with six Turbofan engines.

AST-AVIASPESTRANS CONSORTIUM (RUSSIA)

This a consortium formed in 1989 at Moscow by Russia's Arctic and Antarctic scientific Research Institute for civil aviation, Gasprom Gas Enterprises, Institute of Ocean Geology Engineering Center, Peoples of the North Foundation, Myasishchev Bureau and Promstroy bank. These have a common interest in air transport infrastructures in remote regions of the north, Siberia and Far East. Aviaspetstrans markets services such as monitoring systems and navigation aids as well as aircraft. The first project was the amphibious twin turbo-shaft Yamal capable of autonomous operation in regions with minimal transport infrastructure.

AVIATION SCOTLAND

This company was formed to resume production of and further develop former Island resources. Aircraft ARV-1 Super 2; design and manufacturing rights purchased in 1991; Manufacturing plant and machinery transferred to freehold premises in Burnbnk. The ARV-1 Super 2 is a two seat monoplane.

AVIAT [USA]

They are located at Afton, Wyoming. The Aviat evolved from Pitts Aerobatics company plus manufacture and marketing rights of Pitts Special aircraft acquired by Christian in November 1983. The Pitts Aerobatics factory at Afton became headquarters of Christian Industries was acquired in Afton 1991 by Aviat. Aviat owns production and type certification for the Christian range. Aviat is a wholly owned by White International LTD of Guernsey, Channel Islands. They produce the A-1 Husky, a two seat utility aircraft; the Husky Aviat Acro Husky; and Pitts S-1T Special, a single seat acrobatic airplane; The Pitts S-2B, S-2S and Super Stinker. All the Pitts airplanes are in the acrobatic category.

AVIATICA [RUSSIA]

Aviatika is a Joint Stock Company, established in 1991 by Moscow Undustrial Aviation Association named after Dementyev [MIAA] which is also responsible for MIG production. They are developing single and two-seat light utility multi-purpose aircraft. It is similar to other Russian light aircraft development. The Aviatika 890U is a two-seat biplane. It has non-retractable landing gear. The Aviatika 890 Farmer is a bi-wing agricultural spraying aircraft and the Aviatika 900 is an acrobatic competition aircraft. They all have single engine piston engines and two or three blade propellers.

BEECH AIRCRAFT [USA]

Beech Aircraft Corporation, which is now a subsidiary of the Raytheon Company, was founded in 1932 by Walter H. and Olive Ann Beech at Wichita, Kansas. Raytheon purchased it in 1980,but Beech continues to operate separately, building Civil and Military aircraft, missile targets, and components for aircraft (including control surfaces for Boeing 737) and missiles. The Salina, KS division supplies all wings, nonmetallic interior components, ventral fins, nose cones and tail cones used in the Wichita plant and builds major subassemblies for the Beechjet. The Wichita plant products include subcontracted and metal winglets and composites, and for the McDonnell Douglas landing gear doors for the C-17.

Wholly owned subsidiaries include Beech Aerospace Services, Inc. of Madison Mississippi, which provide worldwide logistic support for Army/Air Force/Navy C-12, Army U-21 and Beech MQM targets, and U.S. Navy T-34C and T-44 trainers in the United States; Beech Acceptance Corporation, which provides business aircraft retail financing and leasing; Travel Air Insurance Company, LTD Bermuda providing liability insurance; United Beechcraft, Inc. providing marketing support for the parent Company. Wholly owned sales outlets include United Beechcraft Inc. of Farmingdale NY and Bedford, Mass. Birmingham AL and Hendrick Beechcraft in Houston, Dallas, Corpus Christi, and San Antonio, TX.

Beechcraft has over 9700 employees worldwide and owns and has 4 million Square feet at Wichita and Salina, KS plus a two story, 100,000 square- foot Corporate headquarters building just North of its flight and delivery complex. The Wichita complex employs about 5700 personnel and the Salina plant employs about 600. Beech is a leading producer of business, corporate and military airplanes, including Jets as well as entering the service arena for the U.S. Military. Beech Aircraft produced more than 7400 aircraft for Allied Air Forces during World War II. Its relationship continued into the early 1950's when Beech was tasked to overhaul 900 of its war era C-45 "Expeditors" for use as administrative and light cargo aircraft and re-designated as the C-45G and C-45H.

BELL HELICOPTER [USA]

Bell Helicopter is one of the first companies of its kind in the United States. It was organized before 1950 and since then has produced more helicopters than any other company. The first was the model 47 with an opposed cylinder gasoline engine and fiberglass blades. These were sold to the military as trainers, rescue and observation helicopters and commercially for utility and civil uses. It had a tail rotor for directional control. They were on of the first to develop turbine engine helicopters, beginning with the UH-1 series. The famous "Huey" was the workhorse of the Military and the AH-1 was the first attack helicopter. The smaller OH-58 was a good seller also, both as an observation platform, and as a trainer. This OH-58 was outfitted with a mast-mounted sight, providing observation without exposing the airframe. Later, the TH-67 Creek was developed for training. The Osprey V-22 tilt-wing multi mission aircraft was developed in conjunction with Boeing.

The headquarters is at Ft. Worth, TX. From 1970 to 1981 Bell Helicopter Textron was an unincorporated division of Textron. Bell became wholly owned by Textron in January 1982. More than 34,000 helicopters have been manufactured worldwide by Bell, with over 10,000 commercial models. Co-production was established in China, Canada, Bell-Agusta in Italy and Fuji in Japan, and South Korea. Agreements were also made with Maquinarias Mendoza CA and Aerotecnica established in Caracas for marketing and support. Bell Helicopter Asia is a fully owned

Singapore based company for marketing and support in Southeast Asia. Most Americans are familiar with the Bell helicopters and the Internet has all the info on each helicopter and the history of their performance worldwide.

BELL HELICOPTER TEXTRON CANADA [Canada]

Bell Helicopter of Canada was established in 1983 at Mirabel, Quebec to manufacture the Bell 206B-3 Jet Ranger III, sold to the U.S. Army as TH-67 Creek. It was very popular with Army Officials. The Canadian plant encompasses over 400,000 square feet of space and employees over 1300 persons. Bell 206-3 production was transferred to Canada in 1886 The TH-67 Creek. It has a two-blade teetering main rotor and a two-blade tail rotor. Options include auto-stabilizer, autopilot and IFR systems. The main rotor blades have extruded aluminum D-section leading edge with honeycomb core behind and covered by bonded skin. The tail rotor has bonded skin without honeycomb core. It is powered by one 420 shp Allison turbojet engine with a rating of 317shp. It has a rupture resistant fuel tank below and behind the rear passenger seat with a capacity of 91 gallons. Accommodations include two seats side by side in front and a three-seat rear bench seat. Main rotor diameter is 33 feet 4 inches, tail rotor diameter is 5 feet 5 inches and overall length with rotors turning is 38 feet 9 ½ inches and the fuselage including the tailskid is 31 feet 2 inches long. Empty weight is 1625 pounds, maximum payload 1400-1500 pounds and maximum takeoff weight is 320 to 3350 pounds. Never exceed speed is 122 knots or 140mph at sea level and service ceiling is 13,500 feet. Hover OGE is 8800 feet and range with maximum fuel and no reserves at sea level is 365 nautical miles or 420 miles, and at 5,000 feet 395 nautical miles or 455 miles.

206L-4 Long Ranger IV is a seven-seat single turbine general-purpose light helicopter. It is a stretched Jet ranger and Canadian production began in 1987. One version is a twin version developed by Tridair and Soloy and described under the Tridair heading. The 206L-4 has the cabin extended to allow club seating and an extra window. Flying controls include autopilot, stabilization and holds for heading, height and approach . It is powered by one Allison turbo-shaft engine rated at 557 shp, transmission rated at 490shp for takeoff and continuous rating of 370shp. It has a rupture resistant fuel tank holding 110.7 gallons. Optional kits include emergency floatation gear, 2000-pound cargo hook, rescue hoist. searchlight requiring high skid gear, and engine bleed air ECS. Main rotor diameter is 37 feet, tail rotor diameter is 5 feet 5 inches and overall length with rotors turning is 42 feet 8 ½ inches. The fuselage including the tailskid is 32 feet 2 ½ inches long. Empty weight is 2274 pounds, maximum internal load is 2,000 pounds and maximum takeoff with normal weight is 4450 pounds and with external load is 4550 pounds. Never exceed speed is 130 knots or 150mph at sea level, and at 5000 feet altitude 133 knots or153mph. Maximum cruising speed at sea level is 110 knots or 126mph, and at 5000 feet altitude, 111 knots or 128mph. Service ceiling at maximum cruise power is 10,000 feet. Hover OGE is 6500 feet and range with maximum fuel and no reserves at sea level is 321 nautical miles or 369 miles, and at 5,000 feet it is 357 nautical miles or 411 miles.

206L Twin Ranger is a seven-seat twin turbo-shaft light helicopter. It is a derivative of the twin engine Long Ranger. Dimensions are the same as the Long Ranger except for modification of cowling contours does the twin-engine installation. The power plant is two Allison turbo-shaft engines rated at 450 shp for 5 minutes for takeoff and 380shp for maximum continuous. Fuel capacity is 112.7 gallons. Empty weight is 2748 pounds, and maximum cruising speed at sea level is 106 knots or 122mph. Economical cruising speed at sea level is 108 knots or 124mph. Service ceiling is 10,000 feet and hover OGE is 6900 feet. Range at sea level with maximum fuel and long range cruising speed and no reserves is 250 nautical miles or 288 miles.

212 with Twin engines, designated as UH-1N. Manufacturing

was transferred to Canada in 1988. The Canadian Forces purchased 50 UH-1N in 1971 and 1972 from the Army and Bell at Ft. Worth, TX. It is used by US Navy, Air Force and Marines. It is powered by two PT6T twin-packs, generating 1800shp for takeoff and 1600shp continuous, manufactured by Pratt &Whitney of Canada. The engines are coupled into a combining gearbox with a single output shaft. It is heated and ventilated. The 212 accommodates a pilot and up to 14 passengers. The cargo version has a capacity of 400 pounds. Optional equipment includes a stretcher kit, cargo hook, cargo sling and rescue hoist. Fuel capacity is 216 gallons and auxiliary tanks provide an additional 180 gallons, for a total of 396 gallons. The main rotor diameter is 48 feet 2   inches, tail rotor is 8 feet 6 inches in diameter and fuselage is 42 feet 4   inches long. Empty weight for VFR is 6097 pounds and for IFR 6277 pounds. Maximum external load is 5,000 pounds and maximum takeoff weight is 11,200 pounds. Never exceed speed and maximum cruising speed at sea level is 111 knots or 128mph. Long range cruising speed at 5,000 feet is 104 knots or 120mph. Service ceiling is 13,000 feet and maximum altitude for takeoff and landing is 4700 feet. Maximum range with standard fuel at 5,000 feet and long range cruising speed and no reserves is 243 nautical miles or 280 miles.

412HP is a four-blade twin-engine utility helicopter. The production was transferred from Ft. Worth to Canada in January 1981. It is an upgrade of the 212 with different versions. The 412SP is a special performance version with increased maximum takeoff weight, new seating options and 55 percent greater fuel capacity. The military 412 is fitted with a chin turret and head tracker helmet sight similar to the AH-1S. The turret carries 875 rounds and can be removed in less than 30 minutes. Other armament includes twin 7.62mm gun pods, single FN Herstal 0.50 in the pod, pods of 17 or 19 2.75 inch rockets, pintle-mounted door guns, four-round 70 mm rocket launcher and a 0.50 inch gun or two 20mm cannon pods. Power is provided by two Pratt & Whitney T6T-3D twin-pack engines with an output of 1910 shp for takeoff and maximum continuous operation. In event of engine failure, the remaining engine can deliver 1140shp for 21/2 minutes or 970shp for 30 minutes. Seven interconnected rupture-resistant fuel cells with automatic shut-off valves have a capacity of 330 gallons. Fuel capacity can be increased to 494 gallons with the addition of auxiliary tanks. Fuselage is 42 feet 4 inches long.

Empty weight is 6654 pounds for VFR and 6759 pounds for IFR. Maximum external hook load is 4500 pounds, maximum takeoff and landing weight is 11,9000 pounds. Maximum cruising speed is 122 knots or 140 mph at sea level, and 124 knots or 143 mph at 5,000 feet. Long-range cruising speed at 5000 feet is 130 knots or 150 mph. Service ceiling is 6800 feet with30 minute power rating, hover OGE is 5200 feet and range at 5000 feet, Long range cruising speed, standard fuel with no reserves is 402 nautical miles or 463 miles. Customers include Venezuelan Air Force, Botswana Defense Force, Sri Lanka, Nigeria, Mexico, South Korea, Honduras, Norway and the Canadian forces.

Bell 230 is a twin turbo-shaft commercial helicopter. Deliveries began in 1992 through the Canadian Government as a repayable loan. Versions are for Utility and executive. Options include dual controls, ECS, auxiliary fuel, force-trim system, more comprehensive nav/com 500 pound capacity rescue hoist, 2800 pound cargo hook, emergency location gear, heated windscreen, particle separator and snow baffles. It is powered by two Allison turbo-shaft engines, each rated at 700 shp for 5 minutes and 622shp for continuous, 779shp for 21/2 minutes and 742shp OEI for 30 minutes. It is also equipped for single engine operation. Fuel capacity is 247 gallons in the skid gear version and 187 gallons in the wheeled version. Optional fuel, 48 gallons, in auxiliary tanks is available. Accommodations are provided for seating 9 persons including the pilots. Options include eight-seat executive, six seat executive or 10 seat utility. Emergency Medical Services versions are available, configured for pilot only plus one or two pivoting stretchers and medical attendants. The entire interior is ram air ventilated and soundproofed, and dual controls are available. Main rotor diameter is 42 feet, tail rotor diameter is 6 feet 10 ½ inches and length of the fuselage is 42 feet   inch for skid gear version and 42 feet   3/4 inches for the wheel gear version. Empty weight is 5,000 pounds with skid gear and 5097 pounds with wheel gear. Maximum external sling load is 2800 pounds, and maximum takeoff weight is 8400 pounds. Maximum cruising speed at sea level for skid gear is 137 knots or 158mph, with wheel gear and standard fuel 141 knots or 162mph. With wheel gear and auxiliary fuel maximum cruise is 137 knots or 158 mph. Service ceiling is 7700 feet and range at maximum cruising speed at sea level, economical cruise, standard fuel and no reserves and skid gear is 385 nautical miles or 443 miles. With wheel gear range is 301 nautical miles or 346 miles. Range at sea level, economical cruise, auxiliary fuel, no reserves and wheel gear is 379 nautical miles or 436 miles.

Advanced developments in Canada are Bell 430, 442 and the Light Helicopter. Bell 430 is a four-blade rotor, higher powered and stretched version of Bell 230. It has Bell 680 all composite rotor blades and a bearing free main rotor. The Bell 230 fuselage is lengthened by 18inches, has about 10 percent power increase through use of a 780 shp Allison turbo-shaft and Takeoff weight is 8600 pounds. The cockpit has LCD integrated instrument display system, Automatic Flight Control System with optional GPS. Bell 442 is a twin turbo-shaft medium weight helicopter, intended to be a Bell 412 successor and international partners are being sought. It is planned to be powered by two Allison/Garrett CTS800 or MTU/Turbomeca/Rolls Royce MTR turbo-shaft engines and a four blade main rotor. The Bell Light Helicopter is planned to be the replacement in the 4,000 to 5000 pound gross weight class and the replacement for the Bell Jet Ranger. It is planned to have a four blade rotor with single or double engines. Also international partners are being sought.

BERIEV [TANTK] RUSSIA

This company was founded by George Mikhailovich in 1932,

except during the second World War from 1942 to 1945. It has been based in Taganrog in the northeast corner of the Sea Of Azov since 1948 and was designated as TANTK in 1990. It now includes the experimental design bureau which has experimental production facilities, a flight test complex, economic, financial and logistics support services with test bases and proving grounds at the Black Sea and Sea of Aznov.

Their main specialty is transports, ranging from the BE-32 carrying a crew of two and accommodations for 14 to 17 passengers, or 17 troops, and ambulance for nine stretchers, six seated casualties and attendant to a crew of eight, 37 or 40 and up to 75 passengers. Military versions are used for cargo transport, surveillance, search and rescue, amphibian, early warning and control and AEW&C versions.

Be-30 is mainly used as a commercial and military transport. It is powered by two 1011shp Mars [Omsk] TVD-10B turboprops driving AB-24AN propellers. Garrett engines are optional. Six internal fuel tanks hold 594 gallons. The civil version accommodates up to 17 passengers and a crew of two. Wingspan is 55 feet 9 1/4 inches and length overall is 51 feet 6inches. Cabin and flight deck are heated and ventilated with portable oxygen bottles, masks, and smoke protection goggles. The military version as described above as troop transport. The NATO name is Cuff. Empty weight is 10,495 pounds, maximum payload is 4190 pounds, maximum fuel is 3750 pounds and maximum takeoff weight is 16,090 pounds. Maximum landing weight is 14,990 pounds. Maximum cruising speed at 9840 feet altitude is Mach .43, 237 knots or 273mph. Takeoff run is 1970 feet and landing run is 2035 feet. Range with 17 passengers is 373 miles, with 14 passengers 596 miles and with 7 passengers, 1087 miles .

Beriev A-40 Albatross, NATO reporting name Mermaid, is a twin turbofan maritime patrol amphibian. It carries extensive avionics and operational systems in primary anti-submarine warfare form. Be-40P is a passenger configuration, Be-40PT is a cargo-passenger, and the Be-42 is a search and rescue version.

They have wingspans of 136 feet 6 ½ inches and length overall is 143 feet 10 inches and Fuselage is 127 feet 8 inches long Maximum takeoff weight is 189,595 pounds, maximum landing weight on land is 160,935 pounds and on water 187,390 pounds. Maximum speed is Mach 0.79. They cruise at 19,700 feet altitude 388 knots or 447mph. They are powered by two Aviadvigatel D-30KPV turbofans, pylon mounted above wing root pods with outward-toed exhaust, each rated at 26,455 lb st. RD60K booster turbojet is rated at 5510lb st in fairing on each turbofan pylon slightly aft and slightly inboard of D-30KPV nozzle with vertically split eyelid jet pipe closure at rear. Wing fuel tanks hold 9,272 pounds and they have a flight refueling probe above the nose. They carry a crew of eight. Range is 2547 miles with maximum payload and 3417 miles with maximum fuel.

Be-42 is a twin turbofan search and rescue amphibian with special SAR equipment and no booster turbojets and no wingtip ESM containers. It caries a crew of nine with provisions for 54 survivors. It is equipped with flares, power boats, life rafts, and onboard equipment to combat hypothermia for 20 survivors, defibrillator, electrocardiograph and other resuscitation and surgical equipment and medicines. It also has electro-optical sensors and searchlights to detect shipwreck survivors by day or night.

Be-200 is a twin turbofan amphibian with the primary mission of firefighting. It has provisions for a crew of two, two cabin attendants and up to 68 tourist class passengers as a passenger airplane. As a Cargo airplane, it can carry a payload of 17,635 pounds. Estimated range is 685 miles with 15,430 pounds of payload. As an ambulance it carries a crew of two, and seven medical personnel, and 30 stretchers in three tiers. It is powered by two D-436T turbofans, each producing 16,550 lb.st. It has much reduced fuel capacity as compared to the A-40. The wingspan is 104 feet, 7   inches and length overall is 105 feet 1 inches. Maximum takeoff weight is 79,365 pounds and maximum landing weight on land or water is 77,160 pounds. Mach number is 0.69 in level flight and maximum cruise is 377 knots or 435mph. Service ceiling is 6,090 feet and landing requirement is 3445 on land and 3610 feet on water. Range with 8818 pounds of cargo is 1133 nautical miles or 1305 miles. With maximum fuel the range is 2485 miles.

Beriev Be-103 is a twin-engine, light business amphibian. It is a low wing monoplane with pneumatically retractable tricycle landing gear. It is powered by two 173 hp. Bakanov M-17 piston engines, each driving a AV-103 three bladed tractor reversible pitch propeller. It can accommodate a pilot and five passengers. The interior is heated and ventilated. Wingspan is 41 feet 9 inches and overall length is 34 feet 3 ½ inches. Weight empty is 2668 pounds, maximum payload including fuel is 1212 pounds. Maximum fuel is 705 pounds and maximum takeoff and landing weight is 3880 pounds. Maximum cruising speed is 143 knots or 165mph. It requires a takeoff runway length of 1115 feet on land and 1640 feet on water and landing 1805 on land and 1805 feet on water. Range with maximum payload is 270 nm or 310 miles and with maximum fuel is 615 miles.

Beriev A-50 is a derivative of and operational characteristics are about the same as the Ilyushin Il-76. It is a four turbine, early warning and control aircraft with NATO reporting name MAINSTAY. It operates with MIG-29, MIG 31 and Su-27 counter-air fighters. It carries a crew of 15 and normally operates on figure eight course at 33,000 feet with 62 miles between centers of two orbits.

BOEING AIRPLANE COMPANY [USA]

The Boeing Company, founded in 1916, is headquartered in Chicago, IL. They employ 150,000 people across the United States and in 70 countries, with major operations in the Puget Sound area of Washington, southern California, and St Louis, MO. Total revenues for 2006 were $61.5billion. Operating units of Boeing are: Boeing Commercial Airplane group and the Boeing Defense and

Space Group. The Defense and space groups the following component divisions: Boeing Computer Services; Electronic Systems Division; Helicopters Division; Military Airplanes Division and Product Support Division. Boeing is the world's leading aerospace company and the largest manufacturer of commercial jetliners and military aircraft combined. Also Boeing designs and manufactures rotorcraft, electronic and defense systems, missiles, satellites, launch vehicles and advanced information and communication systems. As a major service provider to NASA, Boeing operates the Space Shuttle and International Space Station. The company also provides numerous military and commercial airline support services. Boeing has customers in more than 90 countries and is one of the largest U.S. exporters in terms of sales. Boeing Commercial Airplane Group headquarters is in Renton, WA near Seattle. In 1983, they were organized into three divisions; The Renton division produced the 707 and now produces the 737 and 757; Everett Division produces 747, 767 and 777. The Fabrication division provides manufacturing for other divisions. The Materiel Division, created in 1984 covers purchasing, quality control and vendor supplies. Like all commercial airplane companies, Boeing reduced total workforce by 28,000 by mid-1994, of which about 19,000 were in Seattle area production. The merger with McDonnell Douglas gives the combined company a 70 year of heritage of leadership in commercial aviation. Now, the main commercial products are the 737, 747, 767, and 777 airplanes and the Boeing Business Jet. Development is focused on the 787 Dreamliner expected to be in service in 2008. The company now has 12,000 commercial jetliners in service worldwide, which is about 75 percent of the world fleet. The Boeing Commercial Aviation Services, provides unsurpassed, around the clock technical support to help operators maintain their airplanes in peak operating condition. Commercial Aviation Services offers a full range of world-class engineering, modification, logistics and information services to its global customer base, which includes the world passenger and cargo airlines as well as maintenance, repair, and overhaul facilities. Boeing also trains maintenance and flight crews in the 100 seat and above airliner market through Alteon, the world's largest and most comprehensive provider of airline training. The Boeing Defense And Space Group includes the E-3 Sentry AWACS. It was followed by the 767 AWACS which was sold to Japan at a cost of $408.4 million each.

Boeing purchased Vertol, formerly Piaseki Helicopter Corporation, in 1960. The Piaseki Helicopter company was organized before 1950 by the Piaseki family, was and is still located in Ridley township in Pennsylvania and has a flight test center at Greater Wilmington, Delaware. Their specialty is Tandem Rotor helicopters such as the H-21, H-25, CH-46 and Ch-47. All first helicopters were for search, rescue and utility. These roles have been expanded to cargo, passenger, electronic platforms, attack and other roles not previously thought of for helicopters. The purchase by Boeing expanded the outlook for Piaseki and continues the expansion to electronics and other vertical takeoff and landing roles. The workforce at Boeing-Vertol in Pennsylvania is composed of about 7000 employees.

The Military Airplane Division located at Seattle, WA included the development of surveillance and fighter platforms including the Lockheed/Boeing F-22, the Boeing EX, the Boeing/ Sikorsky RAH 66 Commanche and future programs. Boeing has programs in conjunction with Bell and Sikorsky for development of future applications. On December 15, 1996 the Boeing Company and the McDonnell Douglas Corporation merged in a stock-for-stock transaction.

BRITISH AEROSPACE [UK]

British Aerospace [Bae], of Farnborough, Hampshire was organized in 1977, with the nationalization and merger of British Aircraft Corporation [holdings] Ltd, Hawker Siddeley Aviation, Ltd, Hawker Siddeley Dynamics and Scottish Aviation, Ltd. British Aerospace became private sector public limited company in January 1981; residual HM Government shareholding sold in May 1985; responsibility for business activity passed in 1989 to divisions and subsidiaries, currently employing 88,000 personnel, and resulting in sales of $22 Billion dollars or UK 22 billion pounds.

Principal wholly owned companies for aircraft and avionics only are: Aero International Aerospace Ltd; British Aerospace Airbus Ltd [Airbus wings design and manufacture]; British Aerospace Australia Ltd [electronic systems and space equipment]; British Aerospace Defense Ltd [military aircraft, guided weapon systems, ordnance and support services]; British Aerospace Inc. [US holding company for BAE's North American assets]; British Aerospace Regional Aircraft Ltd [regional carriers]; British Aerospace Sweden AB [Defense sales organization]; British Aerospace Ltd [systems and equipment] LTD Electronic systems and equipment for defense and civil applications]; Jetstream Aircraft Ltd [turboprop aircraft]; and Jetstream Aircraft Inc [marketing and support for BAE turboprop aircraft in the West].

Subsidiary and associate organizations for aircraft and avionics only with BAe interest are: Eurofighter Jagdflugzeug GmbH [Eurofighter 2000] 33 percent]; Euroflag srl [new medium-lift military transport aircraft] 20 percent; Euromissile Dynamics Group [Trigat anti-tank guided weapon system] 33 percent; Panavia Aircraft GmbH [Tornado multi-role combat aircraft] 42.5 percent; Reflectone, Inc. [flight simulators and electronic training systems] 53 percent; SEPECAT SA [Jaguar combat aircraft] 50 percent; and Singapore British Engineering pte, Ltd [marketing BAe defense products] 49 percent.

They are now the 4th largest defense company, No 1 European defense company, and in the Top Ten U.S defense companies. They have major operations across five continents with customers and partners in more than 100 countries. Strong positions are in their six home markets; Australia, Saudi Arabia, South Africa, Sweden, UK and the USA.

CANADAIR [CANADA]

Canadair was organized in 1944, acquired by Bombardier in December 1986 and merged with their parent company to form Canadair Group of Bombardier in August 1988. They now have the Business Aircraft, Regional Aircraft, Amphibious Aircraft, Manufacturing, and Defense System divisions. Between 1944 and 2004 they had manufactured about 4200 airplanes. Besides manufacturing aircraft, they sub-contract to other companies, such as nose barrel units for McDonnell Douglas F/A-18, six major fuselage sub-assemblies for Airbus A330/340, Airbus inboard wing leading-edge assemblies for Bae and rear fuselage sections for Boeing 767. They also repair, overhaul, and produce spares for other aircraft.

The Challenger 600 series [Canadian Forces CC-144 and CE-144A] are twin turbofan business, cargo, and regional transports. The landing gear is tricycle type, hydraulically actuated, with twin wheels and oleo-pneumatic shock absorbers. Main wheels retract inward into wing center section and the nose wheel retracts forward. The nose wheel is steerable and self-centering, multiple disc carbon brakes with anti-skid system. Ground turning radius is 40 feet. It is powered by two GE turbofan engines, each generating 9220 pounds st with automatic power reserve, or 8730 pounds st without APR. Pylons are mounted on the rear fuselage and cascade type fan air thrust reversers. Fuel capacity is 2635 gallons from the fuel tank in the center section, one in each wing, beneath the cabin floor, a tank in the tail cone, and auxiliary tanks. It is powered by two GE turbofan engines. It accommodates a crew of two on the flight deck, with dual controls and blind flying instrumentation. The cabin accommodates a maximum of 19 passengers, including toilet, buffet, bar and wardrobe. The Medevac version can carry up to 7 stretcher patients, with an infant incubator, a full complement of medical staff and comprehensive intensive care equipment, including cardio-pulmonary resuscitation equipment physio-control lifepack comprised of heart defibrillator, ECG and cardioscope, ophthalmo scope, respirators and resuscitators, infant monitor, X-ray viewer, cardio-stimulator, fetal heart monitor and anti shock suit. The entire accommodations are heated, pressurized, and air conditioned. Wingspan is 64 feet 4 inches and overall length is 68 feet 5 inches. Empty weight is 20,735 pounds, empty operating weight is 25,760 pounds, maximum fuel is 17,900 pounds, maximum payload is 5240 pounds and maximum takeoff weight is 45,100 pounds. Maximum landing weight is 36,000 pounds. Maximum cruising speed is 476 knots or 548 mph, normal cruising speed is 459 knots or 529mph, long range cruising speed is 424 knots or 488mph and maximum operating altitude is 41,000 feet. Takeoff field length is 6050 feet, landing distance is 3300 feet. Range with maximum fuel, five passengers and at long-range cruising speed, is 3585 nautical miles or 4125 miles and design g limit is +2.6.

The Canadair Regional Jet series 100 is a twin-turbofan regional transport. The landing gear is hydraulically operated tricycle type. Main units retract inward and have Aircraft Braking System. The nose gear is a steer by wire and ground turning radius is 75 feet. It is powered by two General Electric turbofan engines, each with 9220 pounds st with ZAPR and 8729 pounds st tanks hold pressure, and wing leading edges and engine intake cowls are anti-iced by engine bleed air. The windscreen and cockpit side windows, pitot heads, air data vanes, static sensors are electrically de-iced. An ice detection system is standard. It accommodates a crew of two pilots and one or two cabin attendants. Cabin seats up to 50 passengers in standard configuration. Various versions can seat 15 to 50 seats depending on the version desired by the purchaser. Entire accommodations including the rear baggage compartment are pressurized. Wingspan is 69 feet 7 inches and overall length is 87 feet 10 inches. Internal dimensions are: Length, 48 feet 5 inches; height 6 feet 1 ½ inches; floor area 346 square feet; volume is 2015 cubic feet. Stowage volume in the rear baggage compartment is 314 cubic feet; and wardrobes/bins/under-seat areas, 182 cubic feet. Empty weight is 29,180 pounds, operating weight 30,100 pounds, maximum payload is 12,100 to 13,878 pounds, maximum fuel is 9380 to 14,305 pounds and maximum takeoff weight is 51,000 to 53,00 pounds. Maximum operating speed is mach 0.85, high speed cruising speed is Mach 0.80, 459 knots or 529mph and normal cruising speed is Mach 0.74, 424 knots or 488mph. Maximum operating altitude is 41,000 feet, Takeoff field requires 5265 feet and landing field length is 4725 feet. Range with maximum payload at long range cruising and reserve fuel is 980 to 2141 nautical miles, or 2465 miles, depending on configuration.

Global Express is a long-range, high-speed corporate transport. It features a wide body fuselage with about the same length of the Regional Jet. The wings are swept back 35 degrees. There are two rear mounted engines, BMW Rolls Royce 14,690 pounds st each. It has three passenger configurations, Executive, 18 seat, and 30 seat; a crew rest area, large galley and baggage compartment, a modern flight deck that includes six flat panel displays, heads-up guidance system, side stick controllers and electronic library. Wingspan is 91 feet 10 ½ inches, overall length is 99 feet 1 inch and height is 24 feet 3 inches. Maximum takeoff weight is 91,000 pounds and maximum landing weight is 78,600 pounds. Maximum cruising speed is Mach 0.88, long range cruising speed is Mach 0.85 and long range cruising speed is Mach 0.80. Maximum takeoff requirement is 5540 feet. Range with four crew members and 8 passengers at Mach 0.88 is 6500 nautical miles or 7480 miles, at mach 0.85 it is 6330 nautical miles or 7284 miles and at Mach 0.88 it is 5000 nautical miles or 5754 miles.

Canadair CL-415 is a twin turboprop amphibian, with missions planned for firefighting, maritime search and rescue, and special mission. The landing gear is hydraulically retractable tricycle type with self-centering twin nose wheel, which retracts rearward into the hull and is fully enclosed. Main landing gear retracts into wells in the sides of the hull. It has non retractable stabilizing floats, each mounted near the wingtip on a pylon cantilevered from the wing box structure with breakaway provisions. Power is derived from two 2380 shp Pratt & Whitney Canada turboprop engines on damage resistant mounts. The propellers are Hamilton Standard four blade, constant speed, fully feathering and reversible pitch. Two fuel tanks each of 8 identical flexible cells in the wing spar box have a total capacity of 1531 gallons. It uses a pneumatic/electric intake de-icing system. Accommodations include a crew of two side by side on the flight deck with dual controls. Additional stations are available for maritime patrol/SAR versions for a third cockpit member, mission specialist and two observers. Transport options can accommodate 30 passengers plus toilet, galley, and baggage area. Quick-change interiors are available for utility/paratroop drop up to 14 folding canvas seats in the cabin or other missions the customer desires. The wingspan is 93 feet 11 inches, length is 65feet ½ inch and overall height is 29 feet on land and 22 feet 7 inches on water. Propeller diameter is 13 feet   inch, fuselage clearance is 1 foot 11   inches and water clearance is 4 feet 3 inches, ground clearance is 9 feet 1 inch. Typical operating weight empty is 27,223 or 27,783, depending on configuration, maximum internal fuel is 19,250 pounds, maximum payload is 9177 or 13,500 pounds, again, depending on configuration. Maximum takeoff weight is 43,850 pounds on land, and on land and water 37,850 pounds. Maximum touchdown for water scooping is 36,200 pounds and maximum flying weight after water scooping is 46,000 pounds. Maximum landing weight is 37,000 pounds. Maximum cruising speed at 10,000 feet with a weight of 32,500 pounds is 203 knots or 234mph.

CESSNA [USA]

Clyde V. Cessna in established Cessna Aircraft Company at Wichita Kansas in 1911. It was incorporated in September 1927. In

1984, the former Pawnee and Wallace divisions in Wichita consolidated into the aircraft Division. In 1985, General purchased Cessna Dynamics as a wholly owned subsidiary. Textron acquired them in February 1992. The headquarters remains at Wichita, Kansas. Subsidiaries owned are McCauley Accessory Division at Dayton, Ohio and Cessna Finance Corporation in Wichita. They sold 49 percent interest in Reims Aviation of France to Compagnie Francaise Chaufour Investment [CFCI] in 1989. Reims continues to manufacture Cessna F406 Caravan and holds option to build Cessna single-engine aircraft when Cessna restarts production. Cessna employment is presently 9500 worldwide, with about 8000 at Wichita Manufacturing facilities in the U.S. include Wichita, Kansas, Independence, Missouri and Columbus, Georgia. Nine Citation Service Stations are located throughout the United States. They have over 250 Cessna Pilot Centers throughout the U.S., which have been teaching people to fly for more than 30 years.

Cessna is the world's leading producer of general aviation aircraft. They currently produce six single engine piston models, 172, 172SP, 182, T-182, 206 and T-206, four Caravan models [208, Grand Caravan, Grand Caravan Amphibian Caravan Cargo Master] and nine Citation business jet models (Mustang, CJ1, CJ2, CJ3, Encore, Bravo, XLS, Sovereign, and X). They also, in 2007, selected the engine for the CJ4, and the Aero Club of India ordered 11 Cessna Skyhawks. Cessna has produced over 184,000 aircraft. Over one-half of the general Aviation aircraft flying today are Cessna's. They produced 1239 aircraft in 2006. They have orders for 72 new skyhawks to China and see a strong growth in India and China.

CHINA PEOPLES REPUBLIC [CHINA]

The AVIC [Aviation Industries Of China] is the State owned Government entity to oversee the industry. The aviation industry in China employs a workforce of approximately 560,000. Aircraft factories include: China Nanchang Aircraft Manufacturing Company [CNAMC; Xian Aircraft Company [XAC]; Chengdu

Aircraft Industry Corporation [CAC]; Harbin Aircraft Manufacturing Corporation [HAMC]; Shenyang Aircraft Corporation {SAC]; and Chengdu Aircraft Factory [CAF].

The China Nanchang Aircraft Manufacturing Company [CNAMC] was created in 1951, built 379 CJ-5's under license agreement with Soviet YAK-18s, and in 1960s shared a large production program for J-6 fighter (Chinese variant for Mig-19), J-7 III (Chinese equivalent of the Soviet MIG 21) Fishbed J, and the new J-9 which is assumed to be the equivalent of the Soviet MIG 25. They export to Pakistan, Egypt, Iran, Iraq, Albania, Sra Lanka, Zimbabwe and Myanmar, and Bangladesh. In May McDonnell Douglas submitted license agreements for exporting machine tools to China, to be used wholly dedicated to production of 40 Trunkliner aircraft and related work. Under the Trunkliner [MD-80] program, the Chinese factories were responsible for the production and assembling about 75 percent of the airframe structure and the tools were required to produce parts in support of the planned 10 aircraft per year. The machine tools exported by McDonnell Douglas to China had military and commercial applications. Some of these tools were diverted to a Chinese facility engaged in military production. The tools were shipped to three locations contrary to the license agreement and the CATIC assurances regarding the end use. Six of these machines were diverted to the Nanchang Aircraft factory and the rest were stored in two locations in the port city of Tiajin. By August 1996, about 18 months after the diversion was first reported, all the machine tools were at the Shanghai aviation facility. The Xian Aircraft Company produced the H-6 Badger under license from the Soviet Union. They also produce the Y-7 Transport that accommodates 48 to 52 seats. All the Chinese airlines use this airliner for medium and short distance feeder lines. Boeing cooperation with China began in 1980, and as of 1998 there were about 2000 Boeing airplanes flying worldwide that included major parts built by China. These major parts include 737 vertical fins, horizontal stabilizers, forward access doors, and 747 trailing edge ribs. Xian also cooperates in manufacturing parts for Airbus. Xian produced the JH-7 fighter-bomber, the XAC Y-7 transport aircraft. It is a two-engine turboprop. Variations include military and civil passenger, military and civil Cargo. Chengdu Aircraft Industry Corporation [CAC] was founded in 1958. The main body of the company is production of military aircraft and the sub work is concentrated on commercial parts/components. Products include the Changdu J-7/F-7 fighter, based on the MIG 21, and the latest is Chengdu's J-10 multi-range fighter. The J-10 is based on the cancelled Israeli Lavi lighter. All Chinese aircraft manufacturers were involved in co-production of the McDonnell Douglas MD-80 and MD-90. Twenty air transports were produced in China and twenty were purchased from U.S McDonnell Douglas. They manufacture 757 empennage [horizontal stabilizer], vertical fin, and tail section for Boeing. Chengdu has been involved in manufacturing parts for Airbus as well as maintenance tools.

The Chengdu airframe plant is located 6 nm NW of Chengdu in what the Chinese call the "Land of the heaven." The main plant is based in Sichuan Province and CAC is China's second largest fighter production base. Pakistan is the main customer for fighter aircraft from China. The Harbin Aircraft manufacturing Corporation is a major airframe producer, established in 1952, of amphibian airplanes and helicopters and a partner with other Chinese aircraft companies in the manufacturing of fighter/ attack aircraft. They have more than 16,000 employees and are the sole Chinese builder of helicopters and general aviation aircraft.

The Y-12 is a light multipurpose plane designed and produced by Harbin with 12 seats. Gas turbine engines are provided by Pratt and Whitney of Canada a subsidiary of United Technologies of Hartford, Connecticut. They are a partner with other Chinese Aviation Companies in the development of a 58 or 76 seat transport that has a range of about 3200 kilometers. Currently there are about 200 helicopters operating in China's military fleet.

India, whose army is about half of China, has about 540 helicopters. The latest helicopter is a twin-engine light helicopter. The engines are produced by SMPMC at Zhuzhou as WZ8 and WZ8A. The transmission is manufactured by Dongan Engine Manufacturing Co. at Harbin. Hubs and tail rotor blades are manufactured by the Baoding Propeller Factory. Civil models are for various duties including offshore oilrig support, and air ambulance. It can accommodate four stretchers and two seats, or two stretchers and five seats. Military applications include all three of China's armed forces for transport, commando transport, shipboard communication, and anti-tank. They became part of Xian Aircraft Industrial Group in July 1992. They had a workforce of approximately 4,000, including about 650 engineers and technicians, which part or most were transferred to Xia.

Shenyang Aircraft Corporation has about 30,000 employees and has emerged as China's largest fighter aircraft enterprise since the establishment in 1953. The products produced include the J-8Finback, J-8 I Finback A, JH-8 II Finback B and all-weather jet engine air superiority fighters with secondary ground attack. The customers were China PLA and Iran. The current and future military projects are generally unknown because of the customer security requirements. Their facilities include five large assembly plants and 20 associated buildings, a wind tunnel, and electrical power plant. The Shenyang Engine plant is located at the northeast edge of the Mukden airfield and east of the Shen-Yang arsenal. They own an underground POL fuel storage and at least 14 engine test cells.

They have established a venture with Boeing which is the main component of SAC's civilian aircraft section has two projects thus far, tail sections and airplane doors. They have established a world class production facility, importing all parts and raw materials from American and United Kingdom suppliers in order to maintain the required quality control standards. The cooperation of Shenyang with Boeing and McDonnell Douglas in projects for airplane companies from Israel, Germany, Canada and the United Kingdom form the foundation for SAC not being victim of a market driven condition. SAC provided assistance to the Harbin Aircraft Manufacturing Corporation, and the Xian Aircraft Company for production of the H-6 Badger under a license agreement with the Soviet Union. They also assisted to Nanchang Aircraft for the Q-5 aircraft. Chengdu Aircraft factory production of coach and commercial road vehicles is still the backbone of their operation. They have more recently been placed under MAS jurisdiction they have been producing helicopters, and are one of the three members of the China Helicopter Industry Corporation. These other members are the Harbin Aircraft Manufacturing Corporation and the Helicopter Design and Research Institute. Their workforce is about 6,000, with about 800 engineers and technicians.

Their main Helicopter project was the CAF Z-8. It has a rotor diameter of 62 feet, and the tail rotor is just over 13 feet. It is 75 feet long with rotor turning and width over main gear is 17 feet. It is powered by three Changzhou WZ6 turbo-shaft engines with each providing 1550 shaft horsepower. Two are mounted side by side in front of the main rotor shaft and the other is located aft of the shaft. Mission profiles of the CAF Z-8 include troop transport, ASW/ASV, search and rescue, mine laying/sweeping, aerial survey and firefighting.

DE HAVILLAND [CANADA]

De Havilland was established in 1928 and absorbed by the Hawker Sidley Group in 1961. Ownership was transferred to the Canadian Government in 1974, purchased in 1986 by the Boeing Company and was made a division of Boeing of Canada. Boeing then sold 51 percent to Bombardier and 49 percent to the Government of Ontario in 1992. Support is provided by assistance from Ontario and the Federal government. They have phased back to the small production of the Dash type aircraft.

DHC Dash 8 series 100 and 200 are twin turboprop short-range transports. Series 100 was the initial version with Pratt & Whitney 120A engines, and then was designated as Series 200A with increased performance relative to speed and payload. The 200A increased OEI capability and greater commonality with Series 300. It is the same airframe as 100A and B but with Pratt &Whitney 123C engines, give a 30-knot or 35mph increase in cruising speed, allowing airlines to increase frequencies or operational radius. The Series 200B is the same as 200A, but with Pratt & Whitney 123D engines for full power at higher ambient temperatures. Series 100 series have a retractable tricycle type landing gear with twin wheels on each unit. The nose gear is steer-by-wire and retracts forward. The main units retract rearward into the engine nacelles. Power is supplied by two Pratt & Whitney Canada 2,000shp turboprop engines. Each propeller is four-blade, constant speed and fully feathering aluminum/ glass fiber. Series 200B has 2150shp engines. Series 200A are flat rated for full power up to 26 degrees centigrade. Series 200 maintain power up to 45 degrees and propellers have Beta control. Standard fuel capacity in wing tanks is 835 gallons, and optional auxiliary tank increases fuel to 1506 gallons. They accommodate a crew of two on the flight deck plus a cabin attendant. Dual controls are standard. Standard seating is provided for 37 passengers plus buffet, toilet and large rear stowage area. A wardrobe is located at the front of the cabin in addition to overhead storage and under-seat areas. The alternative 39-passenger version passenger/cargo is an available option. The entire cabin is pressurized and air conditioned. Wingspan is 85 feet, and overall length is 73 feet. The propeller diameter is 13 feet, ground clearance is 3 feet 1 inch and fuselage clearance is 2 feet 6 inches. Operating empty weight is 22,600 pounds for 100A, 100B is 22648 pounds, and 200A and 200B are 22,886 pounds. Maximum standard usable fuel is 5678 pounds and options increase that to 10,244 pounds. Maximum payload is 8400 pounds for 100A, 9352 for 100B and 9114 pounds for 200A and B. Maximum takeoff weight is 34,500 pounds for 100A, and 36,300 pounds for

100A, 200A and B. Maximum landing weight for all is 33,900 pounds. Maximum cruising speed is 265 knots or 305 mph for 100A, 270 knots or 311 mph for 100B, and 300 knots345 mph for the 200 series. Range with standard fuel, IFR reserve and full passenger load is 820 nautical miles or 944 miles. With 6,000-pound payload, range is 1100 nautical miles or 1266 miles.

DHC-8 Dash 8M is a multi mission medium utility airplane. U.S Air Force version is E-9A, which used as a missile range control airplane. It relays telemetry, voice and drone and fighter control data and relays the data and observes range with radar. Avionics package includes a large steered phased array radar in the fuselage side and a surveillance radar in the ventral fin. DHC Dash 8 Series 300 is a stretched twin turboprop regional transport. It has a fuselage extension providing seating for 50 passengers plus second flight attendant, larger galley, galley service door, additional wardrobe, larger lavatory, dual air-conditioning packs and optional APU. The engines are Pratt & Whitney 2500 shp driving four blade propellers.

Wingspan is 90 feet and overall length is 84 feet 3 inches. Empty operating weight is 25,700 pounds, maximum usable fuel is 5678 pounds, maximum payload is 11,500 to 13,800pounds maximum takeoff weight is 41,100 pounds to 43,000 pounds and maximum landing weight is 40,000 to 42,000 pounds. Maximum cruising speed is 287 knots or 330 mph for 300A and 285 knots or 328 mph for 300B. Certificated ceiling is 25,000 feet and service ceiling OEI is 13,500 feet. Takeoff field length is 3560 to 3800 feet and landing field length is 3350 feet to 3450 feet. Range with standard fuel reserve and full passenger load is 830 nautical miles or 955 miles, and with 6000 payload is 870 nautical miles or 1001 miles.

DHC Dash 8 series 400 is a further stretched Dash 8. The changes include accommodations for 70 passengers and new engines. The engines are GE/Textron Lycoming GLC38 and Allison AE 2100 driving six-blade propellers. Wing span is 92 feet 3 inches, and overall length is 103 feet 8inches. Operating empty weight is 34,300 pounds, maximum usable fuel is 11,800 pounds, maximum payload is 16,500 pounds and maximum takeoff weight is 57,000 pounds. Maximum landing weight is 56,500 pounds and maximum cruising speed at 95 percent of maximum takeoff is 350 knots or 403 mph. Service ceiling is 20,400 feet, takeoff field length is 3700 feet and landing length is 4090 feet. Range at maximum cruising speed with 70 passengers including baggage and IFR reserve is 925 nautical miles or 1064 miles.

DASSAULT [FRANCE]

Dassault Aviation is located in Paris and employed about 9800 personnel in 1993. They have produced more than 1100 executive jets, orders keep coming in for Mirages and more than 6500 aircraft have been built since 1945. One of the main products are the Mirage 2000 series and versions keep increasing. Versions include, 2000, a one seat multi-role fighter, the 2000B a two seat trainer counterpart of 2000C, 2000BOB is a two place electronics test bed, 2000C is a standard interceptor, 2000D is a two seat attack version of 2000N without ASMP missile interface and nose pitot but with improved ECM and GPS, 2000E is a multi role fighter for export, 2000ED is a two seat trainer for 2000E, 2000N is a two seat low altitude penetration version to deliver ASMP nuclear stand-off missiles, 2000R is a single seat day/night reconnaissance export version of 2000E but with normal radar nose, and sensor pods, 2000-3 is a private venture upgrade, 2000-5 multi-role upgrade incorporating –3 and –4 improvements, 2000S is an export version of 2000D. Customers include France Military forces, Egypt, India, Peru, Abe Hhabi, Greece, Taiwan and Pakistan. It is a low set thin delta-wing with cambered sections. 58-degree sweep and moderately blended root, area ruled fuselage. It is capable of 9g maneuvers and 270 degree rolls at both supersonic and subsonic speeds and carries four air-to-air missiles. Controls are fly-by-wire with autopilot. High angle of attack help control yaw excursions, and small air brakes are above and below the wings. The landing gear is retractable tricycle type with twin nose wheels. There is a single wheel on each main gear. Nose wheels retract rearward and main units retract inward. Electric-hydraulic operated nose wheel steering plus or minus 45 degrees, and a disconnect allows 360-degree swivel for ground towing. Main wheels have non-skid and graphite disc brakes. A compartment is in the lower rear fuselage for brake parachute, arrester hook and/or chaff/flare dispenser. Power is supplied by one turbofan engine rated at 14,462 pounds st. dry and 21,385 pounds st with afterburning or also the available engine rated 22,046 pounds st. an internal wing fuel tank and a fuselage tank. Total fuel capacity is 1050 gallons Provisions are available for total internal and external fuel capacity up to 2430 gallons. There is a detachable flight refuel probe forward of the cockpit on the starboard side. It has accommodation for pilot only in 200C and two in some versions. The cockpit includes Martin Baker zero/zero ejection seat(s) in an air-conditioned and pressurized cockpit. A pilot-initiated automatic ejection is included in two-seat versions. The canopy is hinged at the rear to open upward and covered with gold film to reduce radar signature. All versions have the most up-to-date avionics, weapons selection, and countermeasures available. Wingspan is 29 feet 11 ½ inches, and overall length is 47 feet 1 inch to 47 feet 9 inches for 2000B and N. Empty weight is 16,534 to 16,755 pounds, maximum internal fuel is 6834 to 6967 pounds, maximum external fuel is 8201 to 8270 pounds, maximum external stores is 13,890 pounds and maximum takeoff weight is 37,480 pounds. Maximum level speed at height is Mach 2.2 and at sea level Mach 1.2. Maximum continuous speed is Mach 2.2 for 2000C and E, and Mach 1.4 for 2000N, D, and S. Minimum speed in stable flight is 100 knots or 115 mph, approach speed is 140 knots or 161 mph and landing speed is 125 knots or 144 mph. Service ceiling is 54,000 feet. Range for 2000-5 is 575 to 2071 miles depending on configuration. G limits are + 13.5 and –9 ultimate and normal +9 and minus 4.5.

Dassault Rafale (Squall) is a two seat French Air Force or a single seat French Navy interceptor, multi role fighter and reconnaissance airplane. Rafale B was originally planned to be a dual control, two-seat version for the French Air Force, weighing 772 pounds more and at a higher cost than the C version. Rafale C is a single seat combat version for the French Air Force. Rafale D is the original configuration from which production was derived. It is named the Rafale Discrete or stealthy generic for the Air Force. Rafale M is a single seat version for carrier operation with a weight penalty of 1345 pounds. Takeoff weight is limited to 36,376 pounds, had 80 percent structure and equipment commonality with the C and a 95 percent systems commonality. Customer are anticipated world wide, but slippages in production have delayed deployment. It has minimum weight and volume structure to hold cost down, thin mid-mounted delta-wing with moving canard and individual fixed kidney shaped intakes without shock cones. It is fully fly-by-wire with side stick controller on the starboard console and small travel throttle. Most of the wing components are made of carbon fiber including elevons, wing slats are made of titanium, spar/fuselage attachment fittings are aluminum-lithium, wing root and tip fairings are Kevlar, canard is mainly super plastic formed and diffusion bonding, fuselage is 50 percent carbon fiber, fuselage side skins of aluminum-lithium alloy, wheel and engine doors are carbon fiber, the fin is primarily carbon fiber with aluminum honeycomb core in the rudder. The landing gear is hydraulic, retractable tricycle type with singles main wheels and twin Nose wheels. All wheels retract forward and carbon brakes are on all three units. It is equipped with a brake chute in a cylinder container at the base of the rudder. The Naval version swivels plus or minus 70 degrees and 360 degrees under tow. The arrester hook is hydraulic on the M and tension-stored on the B and C versions. Power is derived from two turbofans, each rated at 19,950 pounds st dry or 16,400 pounds st with afterburning and in production aircraft this is increased to 19,558 pounds st. Fuel in internal tanks is 1406 gallons and external tanks hold 1742 gallons. Pressure fueling can be accomplished in 7 minutes or 4 minutes for external tanks only. A fixed detachable inflight refueling probe is on all versions. The canopy is one-piece blister canopy hinged to open sideways to starboard. The canopy is gold coated to reduce radar reflection. Avionics and armament are up to date with latest advancements in navigation, armament and electronic countermeasures. The wingspan is 35 feet 9   inches and overall length is 50 feet 2 ½ inches. Empty weight is 19,973 for the D and 21,319. Maximum fuel is 17,637 pounds and maximum ramp weight is 47,399 pounds. Maximum level speed at altitude is Mach 2 and at low level 750 knots or 864 mph. Approach speed is 115 knots or 132 mph, takeoff distance for air defense is 1312 feet and for attack is 1969 feet. Radius of action is 590 nautical miles or 679 miles, and for air-to-air and long range, 1000 nautical miles or 1151 miles. G limits are +9 or –3.6.

Dassault Atlantique 2 is a twin turboprop maritime patrol airplane. It is designed to attack surface and submarine targets, lay mines, transport personnel and freight, protect off shore interests and fly search and rescue missions. It has retractable tricycle landing gear, with twin wheels on each unit. They retract hydraulically, the main retract forward into engine nacelles and the nose gear rearward and have disc brakes and anti-skid units. Power is derived from two 6100 pounds st. turboprop engines, each driving a four-blade, constant speed, metal propeller. fuel capacity is 6108 gallons in four pressure-refueled integral tanks in wings. Normal crew is composed of 10 to 12 consisting of pilot, co-pilot, observer in glazed nose, flight engineer on the flight deck, ESM/ECM/MAD operator, radar-IFF operator, tactical coordinator and two acoustic sensors operators at stations on starboard side of tactical compartment and two optimal observers in beam positions at the rear. Rest/relief compartment is in the center of the fuselage. Equipment consists of more than 100 sonobuoys with launcher and 160 smoke markers and flares. Cameras are mounted in the starboard side of the nose and in the bottom of the rear fuselage. Armament consists of all NATO standard bombs, eight depth chargers, up to eight homing torpedoes, seven advanced torpedoes or two air-to-surface missiles in the non-pressurized lower fuselage. It has four under-wing attachments for external stores including two or four ARMAT or magic missiles, or future air-to-air or air-to-surface missiles or pods.

Wingspan including wing tip pods is 122 feet 9 1/4 inches and overall length is 110 feet 4 inches. Propeller diameter is 16 feet. Empty weight is 56,659 pounds maximum internal load is 7711 pounds and with external weapon load is 7716 pounds, maximum fuel is 40,785 pounds and standard mission takeoff weight for ASW or ASSW is 97,440 pounds and for combined ASW and ASSW is 99,200 pounds. Maximum takeoff weight is 101,850 pounds and maximum landing weight is 101,400 pounds. Maximum level speed is Mach 0.73, at optimum height, 350 knots or 402 mph and at sea level, 320 knots or 368 mph. Stall speed with flaps downs 90 knots or 104 mph. Service ceiling is 30,000 feet, runway takeoff is 6037 feet and landing from 50 feet is 4922 feet. Ferry range with maximum fuel is 4900 nautical miles or 5639 miles and maximum endurance is 18 hours.

Dassault Falcon 50 is a three-turbofan long- range business transport. Accommodations include a crew of two on the flight deck side by side with full dual controls and airline type instruments. There is a third seat aft of the co-pilot. Various cabins allow seating from 8 to 12, depending on customer's desires. As an alternative, the cabin can accommodate up to three stretchers, two doctors, and medical equipment. The rear baggage compartment is pressurized and air-conditioned and has a capacity of 220 pounds which is accessible through a separate door on the port side. It is powered by three turbofans, each producing 3700 pounds st. for takeoff. Two engines are mounted on the sides of the rear fuselage and the third is attached by two top mounts. There is a thrust reverser on the third engine. Integral tanks hold 1529 gallons in the wings and 786 gallons in fuselage tanks. Total fuel capacity is 2315 gallons, with single point refueling. The landing gear is retractable tricycle type with twin wheels on each unit. Main wheels retract inward and nose wheels retract forward. Nose wheels are steerable + or - 60 degrees and + or- 180 degrees for towing. Four-disc brakes are designed for 400 landings with normal braking. Wingspan is 61 feet 10 1/2 inches and overall length is 60 feet 9 inches. Empty weight is 20,170 pounds, maximum payload is 4784 pounds and maximum fuel is 15,520 pounds. Maximum takeoff and ramp weight is 38,800 pounds standard and optional 40,780 pounds, maximum landing weight is 35,715 pounds and maximum mach number is 0.86. Long range cruising speed is Mach 0.75, 430 knots or 495 mph. Maximum operating altitude is 45,000 feet, takeoff distance with 8 passengers and fuel for 3500 nautical miles or 4028 miles is 4480 feet. Landing distance at Mach 0.75, with 8 passengers and 45 minutes long-range fuel for 3500 nautical miles is 4028 feet.

Dassault Falcon 900B is a three-turbofan intercontinental business transport. One variant is the Japan ASDF for maritime surveillance with US search radar, special communications radio, operational control station, UH-25A type search windows and drop hatch for sonobuoys, flares and markers. It has a larger cross-section and cabin length than Falcon 50. It has increased span and area, optimized for Mach 0.84 cruise, sweepback on outer wings, added economy and power and three position air brakes. The speed increase is achieved a by mixer compound engine nozzle tailpipe to mix cold and hot air flows. Customers include Algeria, Australia, France, Gabon, Malaysia, Nigeria, Saudi Arabia, Spain, Syria, and United Arab Emirates. It is powered by three 4750 pounds st. turbofans with a thrust reverser in the center engine. The controls are fully powered with artificial feel and three-position air brakes. Total fuel capacity is 2860 gallons. It accommodates a wide range of options for up to 19 passengers plus a crew of two or three on the flight deck, separated from the cabin by a door. Cabin diameter is 8 feet 2 ½ inches and overall length is 66 feet 3   inches. The cabin is 39 feet ½ inch, width is 7 feet 8   inches and height is 6 feet 1 ½ inches. Empty weight is 22,575   pounds, operating empty weight is 23,248 pounds and maximum payload is 4817 pounds and with maximum fuel, 3053 pounds. Maximum fuel is 19,158 pounds and maximum takeoff weight is 45,500 pounds. Maximum landing weight is 42,000 pounds and normal is 27,000 pounds. Maximum operating speed at sea level is Mach 40.87, 350 knots or 403 mph, between 10,000 and 25,000 feet is Mach 0.84, 370 knots or 425 mph. Maximum cruising speed is 500 knots or 575 mph, economical cruise is Mach 0.75. Stall speed is 79 knots or 91 mph and maximum cruising altitude is 51,000 feet. Takeoff field requirement is 4970 feet, landing length is 2300 feet. Range with maximum payload and IFR reserve is 3460 nautical miles or 3982 miles, at Mach 0.75 with maximum fuel 15 passengers and IFR reserve is 3840 nautical miles or 4419 miles, and with eight passengers 3900 nautical miles or 4488 miles.

Dassault Falcon 2000 is a follow-on to the Falcon series and a trans-continental wide-body business transport. It is the same fuselage cross-section as Falcon 900, but 6 feet 6 inches shorter and wings and tail are swept back as the other Falcon series. It is powered by two turbofan engines, each producing 6000 pounds st. Fuel capacity is 1814 gallons. It accommodates a crew of two on the flight deck and up to 12 passengers. The wingspan is 63 feet 5 inches, overall length is 66 feet 4 ½ inches and height is 22 feet 10   inches. Empty weight is 19,700 pounds, maximum payload is 3064 pounds, maximum fuel is 12,154 pounds, maximum takeoff weight is 35,000 pounds and maximum landing weight is 33,000 pounds. Maximum operating Mach no. is 0.85 to 0.87, maximum operating speed is 350 knots or 403 mph, maximum cruising speed at 39,000 feet altitude is Mach 0.83 to 0.85, and service ceiling is47,000 feet. Takeoff field landing with 8 passengers is 5365 feet and landing field length is 2560 feet. Range with maximum fuel, 8 passengers at sea level is 3000 nautical miles or 3452   miles and g limits are +2.64 and –1.

Dassault Falcon 9000 is a long range, three-engine executive jet. It carries between 8 and 10 passengers over 6000 nautical miles or 6905 miles, takeoff weight of 68,000 pounds with 4606 gallons of fuel. The wingspan is 82 feet 9 inches and overall length is 81 feet 2 inches. Empty weight is 34,187 pounds and maximum Mach no. is 0.90, 370 knots or 426 mph and cruise altitude is 41,000 feet. Field length required is 6,000 feet and range with a crew of four and eight passengers at Mach 0.85, is 6000 nautical miles or 6905 miles.

DORNIER [GERMANY]

Dornier was formed in 1922 and has operated as a GmbH since 1972 in Wessling, Germany and has operated since that time. Besides manufacturing they also make sub-assemblies and in the late 1960's and early 1970's they produced the UH-1D under license with Bell Helicopter and the U.S. Army. The Dornier 228 is the most predominant activity at the present time. The 228-212 is the current production version in Europe. Other versions are: 228-201, an Indian production version; 228 Troop that carries 17, 20 or 22 fully equipped troops. It is adaptable for paratroops with folding seats, lightweight toilet, roller door, military nav/com and loadmaster no toilet; 228 Cargo similar to 228-212 but modified for cargo and 228 is Maritime Patrol and Maritime pollution surveillance and photogrammetric/geo-survey. Customers have responded well. Certifications have been accepted by Bhutan, Canada, India, Japan, Malaysia, Nigeria, Norway, Sweden and Taiwan. A tricycle landing gear is retractable with single wheels on main units and a dual wheel nose unit. Main units retract inward into fuselage fairings and the hydraulically steered nose wheel retracts forward. Bendix carbon brakes are on the main wheels. Power is supplied by two 776 shp turboprops and a four blade, constant speed, fully feathering and reversible pitch propeller. Fuel is contained in wing tanks with total capacity of 630 gallons. It accommodates a crew of one or two and 19 passengers. A baggage compartment in the rear of the cabin and externally accessible has a capacity of 463 pounds. Additional baggage area is in the fuselage nose with separate access has a capacity of 265 pounds. All accommodations are heated and ventilated and air conditioning is optional. Wingspan is 55 feet 8 inches, overall length is 54 feet 4 inches and overall height is 16 feet. Propeller diameter is 8 feet 10 inches and propeller ground clearance is 3 feet 6 inches. Empty weight is 7183 pounds, empty operating weight is 8243 pounds and maximum payload is 4852 pounds. Maximum fuel weight is 4155 pounds, maximum ramp weight is 14,175 pounds, maximum takeoff weight is 14,110 pounds and maximum landing weight is 13,448 pounds. Never exceed speed is 255 knots or 293 mph, maximum operating speed is 223 knots or 256 mph, maximum cruising speed at 15,000 feet altitude and weight of 11,684 pounds is 220 knots or 253 mph and economical cruising speed is 180 knots or 207 mph. Stall speed with flaps up is 73 knots or 80 mph and with flaps down, 69 knots or 80 mph. Service ceiling at 100 feet per minute is 28,000 feet and at 50 feet per minute, is 13,000 feet. Takeoff run is 2200 feet and takeoff to 35 feet is 2605 feet. Accelerate/ stop distance with anti-skid is 2330 feet, landing from 50 feet with maximum landing weight is 1320 feet. Range with 1708 pounds of payload at maximum cruise is 1160 nautical miles or 1335 miles and at maximum range speed is 1320nautical miles or 1519 miles. Pratt & Whitney Canada turboprop engines producing 2180shp for takeoff and a Hartzell six- blade composite propeller with electronic syncro-phasing. All fuel is in wing tanks with a total capacity of 1133 gallons. It accommodates a crew of two and cabin attendants, and main cabin seating for 30 to 33 passengers. The landing gear is retractable tricycle with twin wheels on each unit. The nose gear retracts forward and main units retract into sandwich fairings on the fuselage sides. Wingspan is 68 feet 10 inches, overall length is 69 feet 10 inches and overall height is 23 feet 9 inches. Propeller diameter is 11 feet 6 inches and propeller fuselage clearance is 2 feet 5 inches. Empty operating weight is 19,422 pounds, maximum payload is 7605 pounds, maximum baggage load is 1653 pounds and maximum takeoff weight is 30,071 pounds. Maximum landing weight is 29,167 pounds. Maximum operating speed is Mach 0.59, 355 knots or 388 mph. Required landing field length is 3315 feet and takeoff length is 3610 feet. Range at maximum cruise with 30 passengers, allowance for 100 nautical miles or 115 mile diversion and 45-minute hold is 753 nautical miles or 840 miles at 25,000 feet, and at 31,000 feet, 840 nautical miles or 967 miles.

EMBRAER [BRAZIL]

Empresa Brasileira De Aeromautica. SA is located in Sao Jose dos Campos, SP. It was created in 1969 and began operation in 1970. It has 3,229,170 square feet of manufacturing space and employs about 5700 workers. They manufacture the Embraer/ FMA CBA-123 Vector, the EMB-110, EMB 120 and EMB 145. medium transports and The EMB-201/202 Ipanema which are medium transports. The EMB 201/202 Ipena is a single seat agricultural aircraft. Their newest program is a cooperative program with Piper for the Saratoga and Seneca.

CBA 123 and EMB-110 are twin turboprop airplanes with a wingspan of about 60 feet. the power is derived from Garrett TPE331 or Pratt and Whitney Canada PT6A turboprop Engines.

The –145 is a twin turbofan with engines mounted on each side of the rear fuselage. The engines are Allison AE 3007A turbofans and for the basic trainer 312 and 312H have one 1250shp Pratt and Whitney PT6A engine. The 201 and 202 Ipanema are powered by Textron Lycoming IO 540 flat six engines driving a two-blade [or a three-blade] Hartzel constant speed metal propeller. Optional engine is the IO-550-E with a McCauley two-blade propeller. Fuel in integral tanks is held in each wing which have a capacity of 77.1 gallons of which 69.75 is useable fuel. The most desirable for the transport industry is the EMB 120 that can be configured for 30-passenger layout. It is powered by two Pratt and Whitney 115 turboprop engines and Hamilton Standard 14RF four-blade propellers. It can be converted from passenger to cargo in 50 minutes. The wingspan is 64 feet 10 ½ inches. Empty weight is 15,741 pounds. Maximum useable fuel is 5864 pounds. Maximum takeoff weight is about 26,000 pounds, depending on whether is the 120 25,000 feet is 260/270 knots or 299/311 mph. Service ceiling is 30,000/29,000 feet. Range is 575 to 979 miles.

EMB-145/145ER (extended range) has two Allison AE 3007A turbofans pylon mounted on the rear cone of the fuselage. It can accommodate two pilots, flight observer, cabin attendant and 50 passengers. Carryon baggage stowage and galley are at the front, and toilet and main baggage compartment at the rear of the cabin. The wingspan is 65 feet 9 inches and the fuselage is 91 feet 7 ½ inches. Basic operating weight is 25,540 pounds, useable fuel is 9,374 pounds, and maximum takeoff weight is 42,329 pounds. Maximum landing weight is 41,226 pounds. Maximum operating speed is mach .76 or 368 mph. Maximum cruising speed at 37,000 feet is 410 knots or 472 mph, and long range cruising speed at 32,000 feet is 360 knots or 415 mph. Service ceiling is 37,000 feet and range at 37,000 feet, reserves for 100 nautical miles or 115 miles for diversion, 50 passengers is 921 miles. With maximum fuel and 20 passengers the range is 1600 miles.

EMB-312 Tucano [Brazilian Air Force designation T-27] is a turboprop basic trainer. It is powered by one 750 shp Pratt and Whitney of Canada PT6A-25C turboprop driving a Hartzell three-blade, constant speed, fully feathering reverse-pitch adjustment. Two integral wing tanks in each wing hold 183 gallons. Fuel tanks are lined with anti-detonation plastic foam. Gravity fueling point is on each wing upper surface. About 30 seconds of inverted flight is possible. There are provisions for two under wing ferry tanks with a capacity of 174.4 gallons. With the underwing tanks, the ferry range at 20,000 feet is 2069 miles. The EMB 312His a stretched version of the basic –312.

ALX is for the Brazilian Air Force to patrol the border. It differs from the 312 in that it has a Pratt and Whitney PT6A-68-5 engine powering a Hartzell five-blade, constant-speed, fully feathering, reversible pitch propeller. Other changes include strengthened airframe, cockpit pressurization, re-profiled electrically actuated clamshell canopy, zero/zero ejection seats, air cycle air conditioning, crew anti-G system, single point refueling/ defueling, canopy and propeller de-icing, ventral airbrake, hydraulic steering and all-glass avionics including GPS navigation and TCAS. It is able to cover whole primary and half of advanced training syllabus and fly precision weapons delivery and target towing missions. Other programs include their Light Aircraft Program. These include the EM-720D and EMB-810D under license from Piper. The EM-720D is the Piper PA-32-301 Saratoga and the EM- 810D is the Piper PA-34-220T Seneca III.

GROB [GERMANY]

Grob, was founded in 1971 in Mindeheim, Germany and has built over 3600 airplanes. The full name of the company is Burkhart Grob and Raumfahrt & co., associated with Grob-Werke GmbH & co. KG. They specialize in light aircraft with crews of two and four. Grob G115 is a two seat light airplane. Variations are: G115B was the original powered by a 160hp engine. G115CC1 was a later model powered by a 160hp engine, modified tail, fuel in the wings and other improvements; G115C2 featured a 180 hp engine; G115D1 is a fully aerobatic model with a 180hp engine and intended as a professional trainer and glider tug; G115D2 powered with a 160hp engine and named Heron for the Royal Navy; and G115 Bavarian was a special version for International Aero Club of Florida. It includes fuel moved to the wings, better visibility in the canopy, new instrument panel, modified wing sections and tail surfaces. The current version is G115B.

G115B is a low wing monoplane with conventional flight controls. The landing gear is non-retractable with wheel fairings, steerable nose wheel, cantilever spring suspension, hydraulic brakes and a parking brake. Power is derived from a 160 hp

Textron Lycoming O-320DIA engine with a two-blade constant speed propeller or a -320 engine with a fixed-pitch propeller. Both engines have exhaust silencer, electric starter and oil cooler. Fuel capacity is 37 gallons in two wing tanks with mechanical and electric booster pumps. It accommodates a crew of two side by side under a one piece canopy, dual controls and baggage space behind the seats. Electrical power is 24 volt with alternator and battery. Instruments are adaptable to customer needs, including full IFR capability.

Wingspan is 32 feet 10 inches, overall length is 24 feet 5 inches and overall height is 9 feet. Empty weight is 1433 pounds for the C model and 1455 for the D. Maximum takeoff and landing weight is 2028 pounds. Never exceed speed for the C is 171 knots or 196 mph and for the D is 184 knots or 211 mph. Maximum level speed for the C with a constant speed propeller is 135 knots or 155 mph and for the D it is 146 knots or 168 mph. Maximum cruise speed at sea level with a constant speed propeller and 75 percent power is 130 knots or 149 mph for the C and for the D it is 135 knots or 155 mph. Stall speed for both is 46 knots or 53 mph with flaps down and service ceiling is 16,000 feet. Takeoff run for the C is 788 feet and for the D it is 689 feet. Landing run is 591 feet for both. Range with no reserve fuel for the C is 748 miles and 598 miles for the D. Endurance for the C is 6 hours and 20 minutes and for the D it is 5 hours and 10 minutes. G limits for the C are +4.4/-1.8 and for the D are +6/-3.

Grob 115T Acro is a military and commercial pilot trainer. It is all composite construction by hand without autoclaving, to provide ease of repair. The accommodations provide for two pilots seated side by side with room for helmets and parachutes. Rudder pedals and seats are adjustable, the sliding canopy is fully transparent. It has a 24-volt electrical system with nickel cadmium battery and 70 VA alternator. Full night lighting with instrument lighting, heated pitot tube and alternate static source are all standard. It is powered by one Textron Lycoming 260hp at 2700 rpm, flat six engine. The engine produces 235hp at 2400 rpm with an exhaust noise silencer to lower noise level. The propeller is constant speed, four bladed. Tanks in the wings have a capacity of 74 gallons, with a mechanical pump and electrical booster. The landing gear is retractable tricycle type with electro-hydraulic actuation, oleo dampers, steerable nose wheel and foot powered hydraulic disc brakes. Wingspan is 32 feet 10 inches, overall length is 28 feet 1 inch, and overall height is 8 feet 5 inches. Empty weight is 1874 to 1962 pounds, useful load is 992 pounds and maximum takeoff weight is 2976 pounds. Never exceed speed is 205 knots or 236 mph, maximum cruise at 75 percent power and 8,000 feet is 159 knots or 183 mph and stall speed with flaps and gear down is 54 knots or 62 mph. Service ceiling is 18,000 feet, takeoff run is 1181 feet and landing run is 722 feet. Range with no reserves is 870 nautical miles or 1001 miles, endurance is 3 hours and 30 minutes and maximum g limits are +6/-3.

Grob GF 200 is a four-seat touring airplane. It is all composite construction, the cabin is pressurized and it has an advanced profile with the wing tips upswept, and the center mounted engine includes a rear-mounted propeller through a composite extension shaft. The swept fins are on top and swept rudder above and below the rear fuselage. The buried engine greatly reduces noise. The landing gear is tricycle type, main gear has levered suspension and hydraulically actuated wheel brakes. The main gear retracts inward and the nose gear carries a taxi light and retracts forward. Initial design specified a 270hp flat six turbo-charged inter cooled engine with a three-blade, constant speed, pusher propeller; but later thinking came up with a liquid cooled 350 hp engine driving a five- blade pusher propeller, 5 feet 11 inches in diameter. Standard fuel capacity is 92.5 gallons. It accommodates a pilot plus three passengers in a pressurized cabin, and the electrical system is 28 volt 70 amp and also powers landing gear retraction. Wingspan is 36 feet, fuselage length is 28 feet 7 inches and overall height is 19 feet 10 inches. Payload is 1278 pounds, maximum fuel weight is 555 pounds and maximum takeoff weight is 3527 pounds. Maximum cruising speed is 199 knots or 229 mph. Cruising speed at 75 percent power and 20,000 feet altitude is 181 knots or 208 mph. At 100 percent power at 20,000 feet altitude speed is 227 knots or 261 mph. Takeoff run is 1214 feet and landing run is 1089 feet. Maximum range at 75 percent power at 20,000 feet and 45 minutes reserve fuel is 1090 nautical miles or 1255 miles. Range at sea level, 75 percent power and 45 minute reserve is 977 nautical miles or 1124 miles.

Grob Strato 2C is a high altitude long distance stratosphere and climatic research airplane. It is extra long winged and can fly above 95 percent of atmosphere molecules. The range is capable of observing the North and South Poles from neighboring continents. Tractor propellers and low tail-plane changed to pusher engines and propellers and t-tail to improve air sampling from the forward fuselage. The flight controls are conventional. It has a high tail mounted above the rudder. The airframe is constructed in half-shells in molds and then bonded together. It is powered by two 402 hp liquid cooled flat six engines, and variable pitch feathering five blade propellers through reduction gearboxes, with a diameter of 19 feet 8 inches. The turbocharger has a 3.0 pressure ratio driven by the exhaust of each engine. After driving the turbocharger, exhaust gasses are ducted to the turbine of two-stage centrifugal turbochargers which then feeds compressed fresh intake to intake of engine turbocharger and then to the engine. This power plant is considered the only one capable of operation economically at the planned mission altitudes. It accommodates two pilots side by side and two scientists at consoles facing starboard. A toilet, galley and rest areas provide for 40-hour missions without the need for pressure suits. Scientific payloads are inside a pressurized compartment. Cabin air conditioning and pressurization is provided by cooled air from engine turbo-chargers, and oil cooled engine driven generator provides airplane electrical power. Wingspan is 185 feet 5 inches, fuselage length is 73 feet 6 inches and maximum diameter is 6 feet 6 inches. Empty weight is 12,786 pounds, and maximum takeoff weight is 25,794 pounds. Cruising speed at

78,740 feet altitude is 280 knots or 323 mph and at 59,050 feet altitude it is 186 knots or 214 mph.. Time required to achieve 85,000 at weight of 13,690 pounds is 10 minutes. Maximum operating altitude is 85,300 feet, radius mission with 1763 pounds of payload and 7407 pounds of fuel, including 8 hours at 78,740 feet 2175 miles. Long range endurance mission with 2204 pounds payload and 10,802 pounds of fuel 1ncluding 48 hours at 59,050 feet is 9773 nautical miles or 11,247 miles.

GULFSTREAM [USA]

Gulfstream Aerospace is located at Savannah, Georgia was founded in 1958. In 1978, the original facilities were purchased from Grumman Corporation in 1978. Chrysler purchased the facilities from Mr. Paulson in 1985. Mr. Paulson and Forstmann Little and Company completed repurchase from Chrysler in March 1990. Forstmann Little and Company bought out Paulson in September 1992. Allied Signal AirResearch at Long Beach, California was bought by Gulfstream in 1986. Gulf stream also operates a sub-contracting facility at Oklahoma City, Oklahoma. They then outfitted 17 Gulfstream IVs. Gulfstream was purchased by General Dynamics in June 2001. General Dynamics also acquired the Galaxy and Astra SPX and assigned Gulfstream for marketing, servicing, and support. The Astra SPX was re-branded the Gulfstream G100 and the Galaxy re-branded the Gulfstream G200. They made their debut at the 2001 Paris Air Show. The Gulstream flagships now include the G150, G200, G350, G450, G500 and G550. These are the world's most technologically advanced business jet aircraft. More than one-fourth of Fortune 500 companies operate Gulfstream aircraft.

In 1997, Gulfstream introduced a new era in the history of business aviation with the Gulfstream V. It is powered by two Rolls Royce BR 710 Turbofan engines as well as the certified Gulfstream Enhanced Vision System [EVS]. It can fly nonstop for 6500 nautical miles at speeds up to Mach .885. In 1994, they had over one billion dollars in firm orders. The wingspan is 90 feet 10 inches, overall length is 96 feet, 6 inches. Operating weight empty is 45,500 pounds, maximum payload is 6500 pounds, maximum usable fuel is 38,400 pounds and maximum take-off weight is 85,100 pounds.

The Gulfstream IV was introduced in 1985 and the IV-SP was introduced in 1993. They are powered by two Rolls Royce Mk 611-8 turbofan engines rated at 13,850 pounds thrust, with Target type thrust reversers. Fuel capacity in two integral tanks is 4370 gallons. It has a wingspan of 77 feet, 10 inches. Length overall is 88 feet, 4 inches. The baggage door has a width of 2 feet 11 3/4 inches and passenger door is 3 feet. Weight empty is 35,500 pounds, maximum payload is 400 to 6500 pounds. Maximum operating speed is Mach .88 or 391 mph. The Gulfstream IV SP, USAF designation C-20F/G/H, Swedish designation is TP-102. The IV SP include mission outfit for ASW [Anti Submarine Warfare], medical evacuation, surveillance /reconnaissance, Maritime patrol, Electronic Warfare, priority cargo, and administrative transport.

The Gulfstream fleet today includes the larger cabin, ultra-long-range G-550, ultra long range G-500, the large cabin, long range G-450 business jet, and midrange G-350 with heads-up-display. Also available as optional equipment is the large cabin, and midrange G200 and G-250 aircraft.

HAMILTON AEROSPACE [USA]

Hamilton Aerospace was founded by George Hamilton with the assistance Dr. Leo Windecker in 1979 at San Antonio, Texas. Their concept was to use composites to design Military aircraft and the technology would be beneficial to General Aviation. The concept was to design a full product line. Each product line would use mass production techniques, with all products having quality as the premier guideline. All subsystems, avionics, autopilot, landing gear, etc., other than the composite airframe, would be supplied by proven reputable manufacturers. The first was the A-II Avenger all composite multi-mission fighter. This aircraft would use two 4450lb thrust engines, PW 545D turbofans made by Pratt and Whitney of Canada. Wing span is 31 ft 8 in, overall length 34 ft. 2 in. Maximum takeoff weight is 14,200 pounds. Maximum speed is 550 knots or 633 mph. Range is 2589 miles with an endurance of 5 hours. Crew can be either one or two. The H1 Badger by Hamilton is an all composite twin engine, wing canard. tactical airplane. It was designed for a high thrust-to-weight ratio, a high acceleration rate and high rate of climb. This will result in a higher survivability against small arms and short-range shoulder launched weapons. It has a gross weight of 8000 pounds, empty weight of 3500 pounds, fuel 2500 pounds and weapons 2000 pounds. The length is 35 feet, wingspan is 30 feet and max speed is Mach .92 with an endurance of 4+ hours. Ceiling is 40,000 feet. Mission configurations include Primary

Close Air-to-Ground Troop Support, Long Range Patrol, Forward Air Control and Tactical Scout-Reconnaissance. It is resistant to stall and spin, has a 600-foot turn radius and can perform a 90-degree bank in less than 2 seconds. It has a 2500 mile ferry range, is deployable in C-130 with one hour reassembly time, and is Aircraft Carrier compatible. It has low radar signature air ducting to reduce engine C/S signature and is highly maneuverable. The H1 is powered by two Williams FJ33 or Pratt and Whitney 615 engines. Crew can be configured for one or two. The next project is the HX-322 for General aviation. The objectives are: All composite airframe, graphite and glass with epoxy and aerospace foam; structural integrity, enabling rugged and austere operations; twin jet configuration, versatility to accept a number of engines; and excellent speed and maneuvering performance, while retaining business desirable range. Design strength is 12 g's positive or negative. It has a tandem seating arrangement with front solo capability. Cabin pressurization can be provided if desired. The wing span is 29 feet 4 inches, overall length is 33eet and height is 11 feet. It is envisioned for two engines with 800+

pounds thrust each. Gross weight is 4020 pounds, with empty weight of 2200 pounds. Useful load is 1820 pounds and fuel capacity is 400 gallons. Takeoff distance is 1500 feet, and landing roll is 1100 feet. Maximum velocity is 468 mph and cruise is 400 mph at 30,000 feet. Range is 2400 miles and endurance is 6 hours. Recommended Rotation airspeed is 82 mph, approach speed is 95 mph and stall

Another is Hamilton Aerospace UAV, a pilotless vehicle commonly called a Drone. It is multi-mission that is recoverable. The micro-computer has three modes of operation. The first mode is uplink command and control. The uplink command communicates directly to the command computer, and is used for launch, mission modification and primary recovery. The second mode operation is pre-programmed in the command computer to provide mission-specific control points. With the LRRP-V capability, any conceivable mission or multi-mission could be programmed into the computer. The third mode is the emergency recovery mode. If uplink/downlink should be lost, the computer will, after loss of contact fly a return route for reacquisition or recovery. The military applications include but not limited to electronic intelligence gathering, radar jamming, communication interdiction, FEBA, meteorological data and radar relay. It has a payload of 900 pounds, mission specific hardware can be tailored to the service needs. The engine is WJ33-P7W 615. Maximum payload is 2000 pounds. Wingspan is 29ft, At 2000 feet, sea level climb is 2,000 to 3,000 feet per minute and maximum speed is 400 knots.

HELIBRAS [BRAZIL]

Helicopteros DO Brazil, SA is a subsidiary of Eurocopter France, located at Itajuba, MG. It was formed in 1978 and owned jointly by Grupo Bbueninvest (30 percent) and Eurocopter France (45 percent). Inaugurated in 1980, the plant is 129,167 square feet and they employ about 280 personnel. This program includes assembly of single and twin-engine versions of Esquilo and Fennec. The Esquilo is the civil version of single engine AS350BA and AS350B2 and twin engine AS 355F2 and AS355N. The Fennec is the military version of single engine AS550U2 and AS550A2 and twin engine AS555U2 and AS555A2. Brazilian Air Force designations are: CH-50 (555U2) CH-55 (AS555U2), TH-50 (AS550U2), VH-55 (VIP as 550U2). Brazilian Navy designations are:UH-12 (AS550U2), UH-12B (AS55U2), and U-12B (AS555U2). Brazilian Army designation is: HA-1 (550A2). The HA-1 is equipped for tactical support and reconnaissance. The aero-medical versions include a Doctor, nurse, and two stretcher patients in addition to the pilot. It is equipped with Electrocardiograph, respirator, stretchers, battery operated incubator, oxygen /compressed air cylinders and four-way socket for 115v AC and 12v DC power.

HA-1 can include a 20mm gun and 2.75 in unguided rockets or anti-tank missiles and Heli-TOW sighting system. The UH-12 and UH-12B have two Avibras LM-70/7 pods each containing seven SBAT 70 mm rockets, or two FN Herstal twin 7.62mm MAG machine gun pods and a door mounted MAG pedestal. The Dauphin civilian designation is AS 365N2 and Panther is designated as Military AS565 and HM-1 by the Brazilian Army. NIEVA is a subsidiary of Embraer. in 1980 Factory area is 221,521 square feet and has It is located at Aeroporto, Caixa, Cotucatu, and has a workforce of about 400.

ILYUSHIN [RUSSIA]

This company, organized in 1933, was named after Sergei Vladimirovich Ilyushin who died in February 1977 at the age of 82. Since that time more than 60,000 aircraft have been produced. The most prominent aircraft are for Electronic Intelligence and military applications. The IL-20, NATO designation Coot-A, is a military elint/reconnaissance turboprop airliner. It has four AI-20M turboprop engines producing 4190 shp and each engine driving a AV-681 four blade, reversible pitch propeller. Ten flexible fuel tanks in inboard panel in each wing and an integral tank in the outboard panel hold 6261 gallons. Some IL-18 airliners have additional bag tanks in center section, giving total capacity of 7935 gallons. Wingspan is 122 feet 9 inches, and propeller diameter is 14 feet 9 inches. It has a maximum payload of 29,750 pounds and maximum takeoff weight is 141,100 pounds. Maximum cruising speed is 364 knots or 419 mph. Takeoff run is 4265 feet and landing run is 2790 feet. Range with one-hour reserve maximum fuel is 4140 miles and with maximum payload 2300 miles.

IL-22, NATO name Coot-A, Airborne command post adaptations of the IL-18 transport are operational with CIS Air Forces. A variety of external fairings and antennae can be expected to differ from one aircraft to another, depending on expected missions.

IL-38 is an intermediate range shore-based four turboprop maritime patrol aircraft with a NATO name MAY. It has a basic IL-18 airframe and lengthened fuselage and wings moved forward to balance role equipment and stores. The cabin windows are fewer, has a large under nose Radome, MAD tail sting, wing dihedral three degrees from roots and mean thickness/chord ratio 14 percent. It is powered by four ZMKB Progress/Ivchenko AI-20M turboprops, each generating 4190shp and four-blade reversible pitch metal propeller. Engines are started electrically. Fuel capacity is 7925 gallons in multiple bag fuel tanks in center section and in inboard panel of each wing and integral tank in outboard panel.. Pressure fueling is through a NATO standard filler. Specifications are the same as IL-20 except overall length is 129 feet 10 inches, weight empty is 79,367 pounds, maximum takeoff weight is 140,000 pounds, maximum cruising speed at 27,000 feet is 330 knots or 380 mph. Takeoff run is 4265 feet, and landing run with propeller reversing is 2790 feet. Range with maximum fuel is 3887 nautical miles or 4473 miles. Patrol endurance with maximum fuel is 12 hours.

IL-76, NATO name CANDID, and Indian name GAJARA, has several variations for different missions. IL-76M is designated as CANDID B and is for military application; the IL-76TD is an unarmed version; the IL-76 MF is a stretched version aimed for military and civil applications; the IL-76MDK is to allow cosmonauts to experience weightlessness during training; the IL-76MP is designated for firefighting and the IL-76LL is an engine test bed version. Other variants the AEW7C A-50 known by NATO as MAINSTAY, the Be-976 with rotating radome and wingtip avionic pods but with nose glazing retained, used as radar picket to observe flight tests, and the Il-78, NATO name MIDAS, as a refueling tanker. Customers include Russia, Algeria, Czech Republic, India, Iraqi Airways, Poland, China, United Airlines, Libyan Arab Airlines, South African Airways, Syrian Air, and Cuba. The basic characteristics are as follows. It has a retractable tricycle landing gear. The nose gear has two pairs of wheels side-by-side with central oleo. The main landing gear has two units in tandem, each unit with four wheels on a single axle. The power plant is four Aviagatel D-30 turbofans in individual under-wing pods, each pod on a large forward inclined pylon and with a clamshell thrust reverser. Integral fuel tanks between spars of inner and outer wing panels have a total capacity of 28,922 gallons. It has an APU in left side landing gear fairing for engine starting and to supply power for all aircraft systems on ground making aircraft independent of ground facilities. The wingspan is 165 feet 8 inches and length overall is 152 feet 10 inches. Maximum payload is 88,185 pounds for the 'T' and 110,230 pounds for the "TD." Maximum fuel is 187,037 pounds, maximum takeoff weight is 374,785 for the "T' or 418,875 pounds for the "TD." Maximum level speed is 459 knots or 528 mph. Cruising speed is 405-432 knots or 466-497 mph. Takeoff speed is 114 knots or 131 mph and normal cruising altitude is 29,500-39,370 feet. Takeoff run is 2790 feet for the "T" and 5580 feet for the "TD." Normal range is 2700 nautical miles or 3100 miles. Maximum range with reserve is 4163 miles for the "T" and 4535 for the "TD."

There is also an Airborne Command Post version of the IL-76.

IL-78M is a four-turbofan probe-and-drogue flight refueling tanker. The empty weight is 216,050 pounds. The wing tanks hold 198,412 pounds and fuselage tanks hold 61,728 pounds Maximum takeoff weight on concrete for the –78 is 418,875 pounds and 462,965 for the -78M.Maximum landing weight is 333,995 pounds. Normal cruising speed is 405 knots or 466 mph and maximum takeoff run is 6825 feet. Refueling range ranges from 620 miles to 1553 miles, depending on fuel load. It is powered by four Aviadvigatel D-30KP-2 turbofans, each delivering 26,455 pounds st. and carries a crew of seven.IL-86, NATO name CAMBER is a four- turbine medium-range wide bodied passenger transport.

It is powered by four KKBM NK-86 turbofans, each delivering 28,660 pounds st, on pylons forward of wing leading edges. It also has combined thrust reverser/noise attenuators. It seats up to 350 passengers. Wingspan is 157 feet 8 inches, and fuselage length is 184 feet and inch. Maximum payload is 92,600 pounds, maximum fuel is 194,775 pounds, and maximum takeoff weight is 418,875 to 458,560 pounds, dependant on runway size and type. The landing weight is 385,800 pounds. Normal cruising speed at 30,000-36,000 feet is 486-512 knots or 559-590 mph. Range with 88,185 pound payload is 2235 miles and with maximum fuel 2858 miles. There are reports that design ranges are not being met. A German Airline was quoted a maximum range of 1350 nautical miles or 1550 miles in its sales literature. IL-96 is a four-turbofan wide-body passenger transport. It is powered by Aviadvigatel PS-90A turbofan engines, each producing 35,275 pounds st on pylons forward of wing leading edges. They have thrust reversal, and total fuel capacity is 39,166 gallons. It accommodates a crew of pilot, co-pilot, flight engineer, two seats for additional crew or observer, ten or 12 cabin attendants and all tourist accommodations for 300 passengers. Four independent hydraulic systems are standard with explosion proof fluid at a pressure of 3,000 pounds per square inch. The wingspan is 189 feet 2 inches and overall length is 181 feet 7 inches. Basic operating weight is 257,940 pounds, maximum payload is 88,185 pounds and maximum fuel is 253,311 pounds and maximum takeoff weight is 476,200 pounds. Maximum landing weight is 385,810 pounds. Normal cruising is 33,135 to 39,700 feet is 459 to 486 knots or 528-559 mph. Takeoff length required is 8530 feet, and landing is 6500 feet. Range is 5590 to 6835 miles depending on payload and fuel load.

Il-96M, IL-96MK, IL-96MO and IL-96T are different versions. Basic specifications are the same as IL-96-300 with increased capacity. The versions are powered by four Pratt and Whitney PW2337 Turbofans each with 37,000 pounds st and nacelles provided by Rohr Industries, USA Maximum takeoff weight is 595,238 pounds and normal cruising speed is Mach 0.86. Takeoff runway length of 11,000 feet is required and landing length is 7385 feet. Range is about 6195 nautical miles or 7136 miles.

Ilyushin has also produced a small two to five seat single engine aircraft to be applied to training, basic aerobatics and general aviation. It is a low wing monoplane with non-retractable landing gear. It is powered by one Teledyne Continental 210 horsepower IO-360-ES2B flat six engine and the propeller is a Hartzell BHC-C2YF-1BF/F8459A-8R metal two-blade variable pitch. Fuel and oil systems are suitable for inverted flight. Cabin is ventilated and heated and windscreen is de-misted by a fan heater. The electrical system is 27 volt DC with 1800 watt generator and a 25 AH battery. The wingspan is 34 feet 7 inches and length is 26 feet 3 inches. Empty weight is 1587 pounds for the training version and 1686 pounds for the utility version. Payload is 397 pounds and 970 pounds respectively. Maximum takeoff weight is 2127 and 2888 pounds. Maximum level speed at 9840 feet altitude is 143 knots (165 mph) for training and 140 knots (161 mph) for utility. Maximum endurance is two hours. Takeoff run ranges from 525 feet to 1525 feet, and Landing run is from 1545 to 1640 feet. Maximum range with three passengers is 669 nautical miles or 770 miles.

IL-106 is a four-turbine heavy lift military transport. It has high mounted sweptback wings with winglets, wide body fuselage and landing gear housed in pods on sides of lower fuselage outside of the pressure cell. Also it has all sweptback tail surfaces with horizontal surfaces mounted on tail cone. It is powered by four Samara/Kuznetsov NK-92 turbofans each producing 39,680 pounds st in under-wing pods. Wingspan is 191 feet 11 inches and overall length is 188 feet 11 1/4 inches. The nominal payload is 176,366 pounds. The estimated cruising speed is 442-458 knots or 509-528 mph and service ceiling is 45, 930 feet.IL-112 is a twin turboprop short-range passenger transport. It is a high wing monoplane with T-tail, no wing or tail plane sweep, swept back fin and rudder and pressurized fuselage. The freight version has a rear ramp and larger side doors. It is powered by two 2466shp Klimov TV7-117 turboprop engines. It carries a crew of two and up to 32 passengers. The wingspan is 68 feet 10 inches and overall length is 65 feet 7 ½ inches. The empty weight is 16,755 to 18,078 pounds depending on mission configuration. The maximum payload is 6614 to 7716 pounds and maximum takeoff weight is 24,910 to 28,660, again depending on configuration. The maximum cruising speed is 323 knots or 372 mph Nominal cruising altitude is 22,965-26,250 feet and takeoff run is 985 to 1250 feet. Landing run is 985 to 1150 feet. Range at maximum cruising speed with reserve is 932 to 1802 miles.IL-114 is a twin turboprop short-range passenger and freight transport. The IL-114M is powered by two TV7M-117 turboprops with payload of 15,430 pounds. Approximately 10 percent of airframe by weight is made by using composites The landing gear is retractable tricycle with twin wheels on each unit. All retract forward hydraulically. All wheel doors remain closed except when the gear is retracting or extending. The propeller is a low noise six-blade SV-4CFRP. Fuel capacity in integral fuel tanks, is 1839 gallons. It has an APU in the tail cone. Wingspan is 98 feet, 5 inches and overall length is 88 feet and 2 inches. Propeller diameter is 11 feet 9 inches. Weight empty is 33,070 pounds, maximum payload is 14,330 pounds, maximum fuel is

14,330 pounds and maximum takeoff weight is 50,045 pounds. Nominal cruise speed is 270 knots or 310 mph. Takeoff run on a paved runway is 5085 feet and landing run on paved or unpaved is 4265 feet. Range with reserve and 64 passengers is 621 miles, and range with 3300 pounds of cargo is 2980 miles.

ISRAEL AIRCRAFT INDUSTRIES [ISRAEL]

IAI was established in 1953 as Bedek Aviation and ceased being a unit of the Ministry of Defense and became a Government owned Corporation in 1967. The name was changed to Israel Aircraft Industries in 1967, with the number of divisions reduced from five to four in 1988. The present divisions are: Aircraft, Electronics, Technologies and Bedek Aviation. They are is located at Ben-Gurion Airport, Israel. The Aircraft Division is the Division for manufacture of aircraft, and Bedek is the division for single site military airframe, power plant, systems and accessory service, maintenance and upgrading. The Aircraft Division is the only subject for this discussion.

IAI KFIR [Lion Cub] is a single seat interceptor and attack fighter. The current version is Kfir-C-10. Avionics and equipment based on Lavi TD, include a lookdown/shoot down track-while-scan radar in an enlarged nose radome, astronautics mission armament and display computer, cockpit with HUD, up front control panel and two MFDs and ring laser gyro INS, all linked through a digital database, air-to-air refueling probe and two 449 gallon drop tanks. IAI 1125 ASTRA SP is a twin turbofan business airplane This version has a new interior, upgraded digital autopilot and EFIS, aerodynamic refinements for high altitude performance and range with reserves extended by 72 miles. Astra Galaxy is improved and redesigned. The Astra SP has a wing section high efficiency leading edge sweep 34 degrees inboard, 25 degrees outboard, trailing edge sweep on the outer panels. Flight controls are pushrods and hydraulically operated, the tail plane incidence is controlled by three motors running together to protect against runaway or elevator disconnect. Ailerons can be separated in case of jam, spoiler/lift dumper panels ahead of fowler flaps, dual actuated rudder trim tab, interconnected with leading edge slats, both electrically actuated. The structure is one piece, two spar wing with machined ribs and skin panels attached by four main and five secondary frames. The nose avionics bay door and nose wheel doors are made of Kevlar and Nomex honeycomb, and ailerons, spoilers, inboard leading edges are Kevlar. The windscreen is heated and is made of polycarbonate with external glass layer to resist scratching. The landing gear is hydraulically retractable tricycle type with twin wheels on each unit. Main units retract inward and nose wheels retract forward. Main units are equipped with anti-skid multi-disc brakes. A compressed nitrogen cylinder provides additional power source for emergency extension. The power plant consists of two 3700 pound st each turbofans with hydraulically actuated target type thrust reversers. Standard fuel is contained in an integral tank in the wing center section, two outer-edge, with total capacity of 1297 gallons. Additional fuel can be carried in a 100 gallon removable auxiliary tank in the forward area of the baggage compartment. Single pressure refueling point is in the lower starboard side of the fuselage aft of the wing and a single gravity point in the upper fuselage allows refueling of all tanks from one position. Fuel sequencing is automatic. Accommodations are provided for a crew of two on the flight deck, with a sliding door between the flight deck and cabin, Standard accommodations are for six persons in a pressurized cabin and a toilet in the rear. All six seats are individually adjustable fore and aft laterally, can be swiveled or reclined, and all are fitted with armrests and headrests. Maximum cabin occupants is nine. Standard equipment includes electric windscreen wipers, electric windscreen demisting, cockpit and cabin fire extinguishers, axe, first aid kit, wing ice inspection lights, landing light in each wing root, taxing light inboard of each main wheel door, navigation and strobe lights at the wingtips and tail cone, rotating beacons under the fuselage and on top of the fin, and wingtip/tail plane static wicks. Wingspan is 52 feet 8 inches, overall length is 55 feet 7 inches and height overall is 18feet 2 inches. The cabin length is 22 feet 6 inches, width is 4 feet 11 inches and height is 5 feet 7 inches. Basic empty operating weight is 13,225 pounds, maximum usable fuel is 8692 pounds for standard fuel and 9365 pounds with a long-range fuel tank. Maximum payload is 2775 pounds and maximum ramp weight is 23,650 pounds. Maximum takeoff weight is 23,500 pounds and maximum landing weight is 20,700 pounds. Maximum cruising speed at 35,00 feet altitude, knots or 533 mph. Maximum operating speed sea level to 25,000 feet is 363 knots or 418 mph, and above 25,000 feet Mach 0.855. Stall speed at maximum landing weight with flaps and gear up is 111 knots or 128 mph and with flaps and gear down 92 knots or 106 mph. Service ceiling is 19,000 feet, field landing length is 5250 feet at sea level, ISA and 3200 feet at sea level, ISA + 27 degrees and landing with maximum landing weight is 2720 feet. Range with four passengers and 45 minute fuel reserves is 2814 nautical miles or 3238 miles.

IAI Astro Galaxy is a twin turbofan business and commuter transport. The current version is a four/eight passenger business/executive standard model with option for interior seating of up to 19 passengers for regional transport operation. It is designed for trans-Atlantic range, for example, Paris to New York. It is powered by two 5700 pounds st pylon mounted turbofans on the sides of the fuselage. Fuel capacity is 2080 gallons. The entire accommodations are pressurized as well as the baggage compartment. wingspan is 57 feet, 2 inches, overall length is 63 feet 4 inches and overall height is 21 feet. The cabin is 30 feet 4 inches long, 7 feet 2 inches wide and 6 feet 3 inches high. Empty weight is 17,770 pounds, maximum fuel weight is 13,950 pounds and maximum payload is 4200 pounds. Maximum takeoff weight is 33,450 pounds and maximum landing weight is 27,500 pounds. Maximum operating speed is Mach 0.85, maximum cruising speed is Mach 0.75. Stated in other terms, maximum operation speed is 360 knots or 414 mph, maximum cruising speed is 475 knots or 547 mph and long range cruising speed is 430 knots ore 495 mph. Maximum operating altitude is 45,000 feet, field length is 6030 feet, landing length is 340 feet and range with four passengers and established reserves is 3700 nautical miles or 4259 miles.

INTERAVIA [RUSSIA]

Interavia is located in Moscow and specializes in single and two seat small private and acrobatic aircraft. The main accomplishment is the Interavia I-3. It is a single or two seat aerobatic airplane. It is a low wing monoplane with tapered wings without dihedral or anhedral, non-swept tail surfaces, pointed rudder tip and clear view blister canopy over one or two seats. The landing gear is non-retractable tail wheel type with cantilever main wheel legs. It is powered by one 355 hp. VMKB M-14P nine cylinder air cooled, radial engine driving a two-blade propeller with controllable pitch. It has a wingspan of 26 feet 7 inches and overall length of 22 feet ½ inches. Empty weight is 1675 pounds, takeoff weight is 2118 pounds and maximum takeoff weight of 2343 pounds with two seats and auxiliary fuel. Maximum level speed is 189 knots or 217 mph Normal cruise is at 173 knots or 199 mph. Range is 435 miles and g limits are +10/ and –8.

KAMAN [USA]

Kaman Aerospace Corporation was originated by Charles H. Kaman in 1945, in Bloomfield, Connecticut. They were the leaders in development of intermeshing rotors without the need for a tail rotor for produced in the late 1940s and into the 1950s. At that time there were basically two types of rotor blades, either wood or fiberglass. They eventually developed composite blade technology and provided blades to several helicopter manufacturing companies with this technology. Also, they are a major subcontractor in many aircraft and space programs, including design, tooling, and fabrication of components in metal, metal honeycomb, bonded and composites construction, using techniques such as filament winding and braiding.

The next major project of Kaman was the Seasprite (Navy SH-2) with improvements past the G designation (SH-2G). The SH-2 was designed with the conventional design of a main rotor with a tail rotor for directional control. It was designed for shipboard operation in antisubmarine, utility and observation. It has two turbo-shaft T-700-GE-401 engines mounted on each side of the rotor pylon. Fuel capacity is 476 gallons. It has ship-to-air inflight refueling capability. Standard crew is three, a pilot, Co-pilot, and tactical coordinator or sensor operator. Main Rotor diameter is 44 feet, tail rotor diameter is 8 feet one inch, total length excluding the tail rotor is 40feet, 6 inches, with rotors turning, length is 52 feet, 6 inches. With nose and blades folded, length is 38 feet, 4 inches. It weighs 9200 pounds empty, maximum takeoff weight is 13,400 pounds. Cruising speed is 120 knots (138 mph). Maximum range with two external tanks is 500 miles with endurance of 5 hours.

On the commercial side, the Kaman K-MAX is a single seat external lift intermeshing rotor helicopter, commonly called the "Aerial Truck.." It is powered by one Lycoming T53-17A-1 turbo-shaft engine rated at 1350 shaft horsepower. Length with rotors turning is 52 feet. operating weight empty is 4500 pounds, maximum hook capacity is 6,000 pounds, and maximum fuel is 1541 pounds. Uses include logging, firefighting, agricultural spraying, construction, and surveying. In March 1994, the prototype lifted 6173 pounds. The flying difference between this and the conventional helicopter is in the directional control. With the conventional, any movement through the rudder is instant, whereas this intermeshing blade helicopter is not as instant.

KAMOV [RUSSIA]

Kamov was formed in 1947 and continues the work of Dr. Ing Nikolai Kamov who was a leading designer of rotary wing aircraft. He died in 1973 at age 71 but all Kamov helicopters in service have coaxial contra rotating rotors. Some new developments are using the conventional ingle maim rotor and directional tail rotor. Characteristically, The Russian helicopters are very large compared to other countries.

Kamov Ka-26 is a turbine multi-purpose military helicopter, NATO name Hormone. Ka-25PL, named Hormone A, is a ship-based anti-submarine helicopter equipped with surveillance radar, dipping sonar and armed with one torpedo. The Ka-25T, named by NATO as Hormone B, is a special electronics variant providing over-the-horizon target acquisition for cruise missiles launched from ships on which it is based. It has a larger under nose radome with spherical undersurface. All landing gear can retract upward to minimize interference to emissions, and a spherical fuel container on each side of the fuselage. The Ka-25BShZ is equipped to tow minesweeping gear without sonar. The Ka-25PS, named by NATO as Hormone C, is a search and rescue version with special mission equipment such as a hoist.it has a rotor diameter of 51feet 7 inches. The fuselage length is 32 feet. The empty weight is 10,505 pounds and maximum takeoff weight is 15,873 pounds. Normal cruising speed is 104 knots or 120 mph.

Service ceiling is 11,000 feet and range with reserves with standard fuel is 250 miles, and with external tanks 405 miles.

Ka-27 and Ka-28, NATO names Helix-A and D are also twin turbine multi-purpose helicopters. The design intent is to overcome the inability of the Ka-25 to operate sonar at night and in adverse weather. The Ka-27PL (Helix-D) operates with a crew of three and operate in pairs, one tracking hostile submarines and the other dropping depth charges. The Ka-PS, NATO name Helix-D, is a search and rescue and plane-guard helicopter. It has an external tank on each side of the cabin and a winch beside left cabin door. The Ka-28 is an export version and reported to be TV-

3-117BK turbo-shaft engines and 8113 pounds of fuel in 12 tanks.

Ka-32, has many characteristics of Ka-27 and Ka-28. It has a longer and a more spacious fuselage pod, no central tail fin and a different nose radome. Similar dimensions allow the Ka-27 to stow in shipboard hangers with rotors folded and use deck lifts built for the Ka-25. The equipment includes an infrared jammer at the rear of engine bay fairing, a station keeping light between ESM radome, a chaff dispenser, color coded identification flares, and sonobuoys stowed internally. Otherwise it is externally the same as Ka-32. Cruising speed is 124-129 knots or 143-149 mph. Radius of action against submarine cruising at up to 40 knots at a depth of 1640 feet is 124 miles.

Ka-29 (NATO name Helix-B) is a twin turbine transport and electronic warfare helicopter. The Ka-29RLD is a radar picket helicopter. The rotor diameter is 52 feet 2 inches and overall length is 37 feet 1 inch. Empty weight is 12,170 pounds, normal takeoff weight is 24,250 pounds and maximum is 27,775 pounds. Nominal cruising speed is 127 knots or 146 mph. Service ceiling is 14,100 feet and combat radius is 54 nautical miles or 62 miles. Range with maximum fuel is 285 miles and ferry range is 460 miles.

Ka-32, NATO name Helix-C, is a twin turbine utility helicopter. The Ka-32T is a utility transport, ambulance and flying crane for airways, local routes, including support of offshore drilling rigs. The Ka-32S maritime version has more comprehensive avionics, including under nose radar for IFR operation from icebreakers in adverse weather and over terrain without landmarks. The electrically operated hoist has a capacity of 661 pounds and additional fuel tanks on each side at top of cabin. The –32K is the flying crane version with retractable gondola. The Ka-32A1 is a firefighting version. They are powered with two 2190 shp Klimov TV-3-117V or TV3-117VMA turbo-shaft engines. Rotor diameter is 52 feet 2 inches and overall length is 40 feet 2 inches with rotors folded. Empty weight is 14,330 pounds and normal takeoff weight is 24,250 pounds. Maximum cruising speed is 124 knots or 143 mph, service ceiling is 16,400 feet and range with maximum fuel is 497 miles and endurance with maximum fuel is 4 hours and 30 minutes.

Ka-50 Werewolf has a NATO name of Hokum. It is a twin turbine close support helicopter. Hokum A was limited production and was a basic single-seat close-support helicopter. Hokum B is a two seat trainer and combat aircraft. It was considered the World's first single seat close support helicopter. It has co-axial contra-rotating widely separated three blade rotors with swept retractable landing gear, mid swept wings with ECM pods at tips, four under wing weapon pylons and engine has prominent exhaust heat suppressors. It is highly agile for fast, low flying close attack missions Composite materials constitute about 35 percent by weight of structures, including rotors. About 770 pounds of armor protect the pilot and critical components. Canopy and windscreen panels are bulletproof glass. It is powered by two 2190 shp Klimov TV3-117VK turbo-shaft engines. It has provisions for four under wing 132-gallon auxiliary fuel tanks. It has a specially designed pilot ejection system for safe ejection at any altitude by an explosive separation of rotor blades and cockpit roof. Pilot is ejected from the cockpit by large rocket. As an alternative, the pilot can jettison stores and doors before rolling out of the cockpit sideways. The rotor diameter is 47 feet 7 inches and length with rotors turning is 52 feet 6 inches. Normal takeoff weight is 21,605 pounds and maximum is 23,810 pounds. Maximum speed is shallow dive is 189 knots or 217 mph and in level flight is 167 knots or 193 mph. Estimated combat radius 135 nautical miles or 155 miles. Endurance with 10 minute reserve is 1 hour 40 minutes and with auxiliary tanks is 4 hours. G limit is +3.

Kamov Ka-126 is a turbo-shaft version of the Ka-26, designed for light general purposes.it has a non-retractable landing gear. For passengers, it has a pod similar to the US Sikorsky flying crane concept, for up to seven passengers, ambulance pod accommodates two stretcher patients, or two seated casualties and medical attendant. It is powered by one 700 shp Mars (Omsk) TV-O-100 turbo-shaft installed centrally in a streamlined fairing above the cabin Fuel capacity is 211.3 gallons contained in two forward and one rear tank. There are provisions for two external tanks on sides of fuselage containing 84.5 gallons. For agricultural purposes, a chemical hopper has a capacity of 264 gallons and dust spreader or spray bars are fitted on the center of gravity. The aircraft can also accommodate either an open platform for freight hauling or hook for slinging loads at the end of a cable or in a cargo net. It has a rotor diameter of 42 feet 7 inches and overall length of 42 feet 7 inches with rotors turning. The passenger pod length is 6 feet 8 inches and width is 3 feet 11 ½ inches. Empty weight is 4222 pounds, maximum sling load is 2205 pounds, maximum internal fuel is 564 pounds, auxiliary fuel is 564 pounds, and maximum takeoff weight is 7165 pounds. Economical cruising speed is 86 knots or 99 mph at sea level with maximum payload is 157 miles, with maximum internal fuel 443 miles, with auxiliary fuel and no reserves, 630 miles. Maximum endurance with internal fuel and no reserves, 5 hours 36 minutes.

Kamov Ka-128 is an upgrade of the Ka-126 is a general purpose light helicopter. The differences are an upgrade of the power plant and addition of an intermediate gearbox. The power plant is one 712shp Turbomeca Arriel 1D1 turbo-shaft with the same rated transmission. These changes the performance and weight but not the other characteristics of the Ka-126.Kamov Ka-226 is a twin turbine utility and agricultural helicopter. It is a refinement of the Ka-26/126 with changes to shape of the nose, twin tail fins and rudders. The passenger pod holds two three-passenger rearward and forward seats facing each other in bench seats. A seventh passenger is accommodated beside the pilot on the flight deck. The engine is two 420shp Allison 250-C20B turbo-shafts side by side rearward from rotor mast in the same position as the single engine Ka-126, with individual drive shafts to rotor gearbox. The transmission rating is 840shp. Fuel capacity is 198 gallons in tanks above and forward of payload module area.

provisions are included for two external tanks on sides of fuselage with a total capacity of 84.5 gallons. The rotor diameter is 42 feet 7 inches and overall length excluding rotors 26 feet 7 inches. Empty weight is 4304 pounds, maximum payload is 2865 pounds, maximum internal fuel is 1322 pounds and auxiliary fuel is 564 pounds. Normal takeoff weight is 6835 pounds and maximum is 7500 pounds. Economical cruising speed is 100 knots or 115 mph and service ceiling is 16,565 feet. Range with maximum payload is 20 nautical miles or 23 miles; with maximum internal fuel 374 miles and with auxiliary fuel no reserves, 542 miles. Maximum endurance is 4 hours and 38 minutes with internal fuel and no reserves.

LEARJET [USA]

The Learjet name has been synonymous of business and corporate aviation since inception in 1960 by Bill Lear, and moved to Wichita, Kansas in 1962 where it was renamed to Lear Jet Corporation. Gates Rubber Corporation bought about 60 percent and in 1967 the name changed to Gates Learjet Corporation. In 1967 64.8 percent of stock was bought by Integrated Acquisition, Inc. and renamed to Learjet Corporation. During 1988, all manufacturing activities were moved from Tucson, Arizona to Wichita, Kansas. They left only customer service and modifications at Tucson. Canada's Bombardier purchased Learjet for 75 million dollars in 1990 and changed the name to Learjet, Inc.

Learjet is a major subcontractor for Martin-Marietta Manned Space Systems, Boeing and U.S. Air Force. Total workforce is about 2800 with the total at all locations about 3700. In 1989 Learjet purchased manufacturing and marketing rights and tooling of Aeronca thrust reversers for Learjet and other aircraft.

The Learjet 35 and 36A were designated by the U.S. Air Force as C-21. The wingspan is 39 feet 6 inches. Length overall is 48 feet 8 inches. Empty weight is 10,119 pounds; maximum takeoff weight is 18,300 pounds; and maximum landing weight is 15,300 pounds. Maximum airspeed is Mach.81, maximum speed at 25,000 feet altitude is 471 knots or 532 mph. Service ceiling is 41,000 feet. They are powered by two Garrett TFE731-2-2B, each rated at 3500st, pod mounted on sides or rear fuselage. Fuel capacity with internal and wingtip tanks is 925 gallons for the 35A and 1104 gallons for the 36A. Range is 2527 to 2902 miles for the 36A. Special mission alterations are: Maritime patrol, Reconnaissance utility and Japan for target towing, antimissile simulation and ECM.

Learjet 31A was unveiled in 1987, with updated technology and improvements in performance. As with all other Learjets, the tricycle retractable tricycle landing gear is standard. Range is 1797 to 2079 miles.

Learjet 45.and 45XR took their first flight in 1992 included all updates as well as seating for 10 to 12. Range with four passengers is 2129 to 2532 miles. Learjet 60 is a medium range business jet. The wingspan is 43 feet nine inches. Length is 58 feet 8 inches. Empty weight 13,840 pounds, maximum fuel capacity is 7850 pounds. Maximum takeoff weight is 22,750 to 23,100 pounds. Maximum cruising speed is 463 knots or 533 mph. Maximum operating altitude is 51,000feet. Range is 2760 to 3153 miles. Engines are two Pratt and Whitney Canada PW305 turbofans, each rated 4600 lb st. Fuel capacity is 7850 pounds.

LOCKHEED [USA]

The Lockheed Corporation is located at Calabasas California. They are the third largest U.S DOD Contractor. They are divided into four operating companies: Lockheed Aeronautical Systems Company; Lockheed Advanced Development Company; Lockheed Fort Worth Company and Lockheed Aircraft Services Company.

The Lockheed Aeronautical Systems Company is located at Marietta, Georgia. Workforce is about 12,000. The Lockheed Advanced Development Company is located at Palmdale,

California. They are specialized development programs for "black" or covert operations. The Lockheed Fort Worth Company located at Fort Worth, Texas. Employment is about 6,000. Lockheed Aircraft Services Company located at Ontario, California is the oldest aircraft modification company in the world. They have designed, fabricated and installed major aircraft structural modifications and integrated complete avionic systems for U.S. military, foreign Governments and commercial customers.

MCDONNELL DOUGLAS [USA]

J.S. [Sandy] McDonnell of St. Louis MO, was Princeton graduate in Physics and MIT in Aeronautical Engineering and learned to fly through the Army Air Service. In 1928 he started his first company to build the "Doodlebug", but since it found no market, he spent the next 10 years working for several companies and finally as Chief Engineer with the Glen L. Martin Aircraft Company. He resigned from Martin in 1938, was determined to form his own company, and in 1939 he incorporated the McDonnell Aircraft Corporation in St Louis, MO. Within the next three decades, the company would become the leading producer of jet fighters and would build the first spacecraft to carry an American into orbit. In the mid-sixties, McDonnell Aircraft was the largest employer in Missouri, and in 1967 it expanded its operation by merging with the largest employer in California, Douglas Aircraft Company. In the same year North American Aviation and Rockwell merged. Before merging with Douglas, they successfully produced the F-2 Banshee, the F-101, and F/A-4 Phantom.

In 1967 McDonnell Douglas supplied the A-4 Skyhawk to the Royal Australian Navy and the DC-9 to Trans-Australia Airlines and Ansett. From 1970 to 1973, the RAF leased 24 F-4 Phantom fighters from the USAF to fill the capability until the F-111 was available. From 1985 to 1990, 73 F/A-18A/B Hornet fighters were assembled in Australia by STA under license to McDonnell

Douglas, with a large number of its components manufactured locally by CAC, ASTA and 15 other companies. In 1996, Rockwell International sold its aerospace and defense business [space systems, missile systems and military aircraft] to Boeing, including most of its Australian operations to form Boeing Australia Limited. In 1998 Boeing Australia Limited moved its headquarters from Sydney, New South Wales to Brisbane, Queensland. Donald Douglas incorporated Douglas Aircraft in Long Beach, CA in 1921. The first contract was to build torpedo bombers (DT-1 and DT-2 versions) for the U.S. Navy. In the subsequent five years, variants of the DT were used as observation biplanes, military transports, cargo carriers, aerial tankers and aeromedical evacuation. Commercial DT aircraft entered service as mail planes (the Douglas M-2) and by 1926, they were producing 120 aircraft per year. In 1936, Douglas Aircraft Company sold their first DC-2 to Australia, followed soon after by the Dakota of World War II fame. Within two years of operation, an estimated 90 percent of world-wide passenger traffic was carried by DC-3. Over 10,000 of these DC-3 transports were produced. During World War II, many Douglas aircraft operated with the Australian Defense Force including the highly successful DC-3, the Dolphin and the Boston The DC-4 entered commercial service in Australia in 1947, the last being retired in 1984. In the U.S. this company grew into a huge conglomerate of transport and cargo aircraft. They absorbed Hughes Helicopters in 1984 and renamed it to McDonnell Douglas Helicopter Company. The Company was renamed the McDonnell Douglas Helicopter Systems. As of last reports, this company employed 70,000 personnel worldwide. Thus, they are in a multi-faced business of Civil aircraft, military aircraft and Helicopter Systems. After a downturn in military and civil business about 1991-1992, they have recovered to be a leading conglomerate throughout the world.

On December 15, 1996 the Boeing Company merged with the McDonnell Douglas Corporation in a stock-for-stock trade.

MiG [RUSSIA]

1n 1971 Artem I. Mikoyan the original head of this OKB in Moscow was succeeded by Rostislav A. Belyakov. He supervised production completion of the MiG-23 and MiG-25, designed by Mikoyan Later aircraft such as the MiG-27 Mig-29, and MiG -31 were designed under Belyakov, developed, and put in service. Many U.S Veterans remember the Mig-17 and MiG-19 from the Korean and Viet Nam wars. The MiG series were all military fighter/Attack Aircraft. Only the later MiG series will be described here. However, they are presently diversifying by developing training and light transport aircraft, and air-cushion landing gear. Also, they are manufacturing horizontal surfaces for Dassault Falcon 900 business jets.

The Mikoyan MiG-23 series, NATO designation as Flogger-A, B, C, E, F, G, H and K, have a wingspan of 45 feet 10 inches fully spread. Fully swept it is 25 feet 6 inches Overall length is 54 feet 10 inches including the nose probe. Empty weight is 22,485 pounds, takeoff weight is 39,250 pounds and maximum internal weapon load is 6615 pounds. Maximum landing weight is 26,015 pounds. Maximum level speed with 72 percent wing sweep is Mach 2.35 or 1553 mph, with 16 degree sweep Mach .88 or 584 mph and with 72 degree sweep Mach 1.10 or 838 mph. Landing speed is 140-151 knots or 162-174 mph and service ceiling is 60,70 feet. Takeoff run 1640 feet and landing run is 2460 feet. Combat range is 715 miles with six air-to-air missiles and with 4410 pounds of bombs, 435 miles. Range with maximum internal load is 1210 miles and with three external tanks 1750 miles G limit below Mach .85 is +8.5 and above Mach .85 it is +7.5.The power plant is one Soyuz/Khachaturov R-35-300 turbojet rated at 28,660 pounds st with maximum afterburner. It has a water injection system with a capacity of 7.4 gallons. Three fuel tanks in fuselage aft of cockpit and six in wings hold 1122 gallons. There are attachments for JATO (Jet Assisted Take Off) on each side of the fuselage aft of the landing gear.

The Mikoyan MiG 27, NATO designation Flogger D and J, is a single seat variable geometry ground attack aircraft. It is based on Mig-23BK. It has one Soyuz/Khachaturov R-29B-300 turbojet rated at 17,626 pounds st with maximum afterburner, two-position (on/off) afterburner nozzle is consistent with the requirement of transonic speed at low altitude. Internal fuel capacity is 1426 gallons with provisions for up to three 209-gallon external tanks. It has a flare dispenser on fuselage on each side of dorsal fin. It is armed with one 23mm twin-barrel gun in fuselage belly pack with 260 rounds. Bomb/JATO rack on each side of rear fuselage, five other pylons for external stores, 8818 pounds of weapons including tactical nuclear bombs, R-3S and R-13M air-to-surface missiles, Kh-23 laser guided radio command guided air-to-surface missiles (laser guided KH-29 on Mig-27D), 240 S-24 rockets, and 22 50 or 100 kg, nine 250 kg, or eight 500 kg bombs in napalm containers. The dimensions are the same as Mig-23, plus these improvements. The overall length is 56 feet inches, empty weight is 26,252 pounds, maximum internal fuel is 10,053 pounds and normal takeoff weight is 39,905 pounds. Maximum takeoff weight is 44,750 pounds for the A to 45,570 pounds for the K. Maximum landing weight is 37,475 pounds. Maximum speed at 26,250 feet altitude is Mach 1.7 or 1170 mph. Service ceiling is 45,900 feet and takeoff run is 3120 feet. Landing run with brakes and chute is 2950 and without brakes and chute 4265 feet. Combat radius with 7 percent reserve and two KH-29 missiles is 140 miles. With two KH-29 missiles and external auxiliary fuel tanks, 335 miles. G limit is +7.0.

The Mikoyan MiG-25, NATO name FOXBAT, is a single seat interceptor, reconnaissance aircraft and two-seat conversion trainer. Variations include MiG-25RB (Foxbat) B) Which is a high altitude reconnaissance bomber; MiG-25RBV and RBT; MiG-25RU (Foxbat) C) which is a training version; MiG-25RBK; MiG-25RBS (Foxbat D); MiG-25PD (Foxbat E) with upgraded R-15BD-300 turbojets with a life of 1000 hours instead of the former 150 hour, And upgraded mission equipment; MiG-25PDS

(Foxbat E) with front fuselage lengthened by 10 inches to accommodate in-flight refueling equipment on some aircraft; MiG-25BM (Foxbat F) Defense suppression variant ECM instead of reconnaissance module in lengthened nose, underbelly fuel tank; and four AS-11 anti-radiation missiles under wing to attack surface-to-air missile radar over standoff ranges. Customers include Algeria, India, Iraq, Libya, and Syria. Engines are two R-15BD-300 single shaft turbojets, each rated at 24,700 pounds st, with afterburning in compartment of silver-coated steel. Water-methanol injection. Fuel is contained in two welded structural tanks occupying 70 percent of fuselage volume. Total fuel capacity is 4665 gallons with provisions for an additional 1400 gallons in underbelly tank. Wingspan is 45 feet 11 inches and overall length is 78 feet 1 inches. Maximum internal fuel is 32,120 pounds for the P series and 33,609 for the R series. Maximum fuel with underbelly tank is 41,755 pounds. Takeoff weight clean with maximum internal fuel for the P series is 76,985 pounds, and with the P series with four R-40 missiles is 80,950 pounds. For the R series, it is normal 81,570 pounds and maximum is 90,830 pounds. Maximum landing weight is 52,910 pounds. Maximum permitted Mach number is 2.83. Maximum level speed at 42,650 feet altitude is 1620 knots or 1865 mph. At sea level it is Mach 0.98, 647 knots or 745 mph. Takeoff speed for the P series is 195 knots or 224 mph. Landing speed for the P series is 157 knots or 180 mph and for the R series 140-151 knots or 161-174 mph. Service ceiling is 67,90 for the P series and 69,900 feet for the R series Takeoff run for the P is 4100 feet and landing run for the P is 2625 feet with brakes and chute. Range for the P at supersonic speed is 776 miles; subsonic, it is 1075 miles with maximum internal fuel. The R series at supersonic speed is 1015 miles and at subsonic speeds 1158 miles. Range with external tank for the R series is 1323 miles at supersonic speeds and 1491 miles at subsonic speeds. Endurance is 2 hours and 5 minutes and g limit is +4.5.

Mikoyan MiG-29, NATO name Fulcrum, is an all-weather single seat counter-air fighter with attack capability, and two-seat combat trainer. The MiG-29 UB (Fulcrum B) is a combat trainer with second seat forward of normal cockpit under continuous canopy, periscope for rear cockpit and with underwing stores retained. The MiG-29S (Fulcrum C) has increased internal fuel capacity of 20 gallons, maximum fuel capacity of 2177 gallons for maximum range of 1800 miles; Radar capable of engaging two targets simultaneously; 8820 pounds of bombs and maximum takeoff weight of 43,430 pounds. The MiG-29SE is an export version of the MiG-29S and has six R-77 AAMs. Maximum takeoff weight is 44,090 pounds and service ceiling is 59,055 feet altitude. The MiG-29M is an advanced fighter for control of upper airspace, ground attack, and navy high altitude precision weapons control. It is a redesigned airframe with two 19,400 pounds st. Klimov RD-33K turbofans, fly by wire controls, and two CRT's for the cockpit, increased volume fuel tanks, eight under-wing hard points for 9920 pounds of stores including four laser guided missiles or TV guided air-to-surface missiles, eight air-to-air missiles or 500 kg of TV guided bombs.MiG-29ME is the export version of the MiG-29M. It has about the same improvements including the airframe, radar and weapons.

The MiG-29K was a carrier-based fighter for the Navy. It has wing folding, strengthened landing gear, tail hook, retractable fueling probe and other Navy requirements. Development of this version was ended because of non-selection by the Navy for carrier deployment. The MiG-29KVP was refined for STOL and carrier operations. The landing gear is a tricycle with dual wheels on the nose wheel. It is very heavily armed with radar and Infrared guided missiles, plus flare dispensers, napalm bombs, rockets for attack role and one 30mm gun is mounted in the left wing root leading edge extension with 150 rounds. The flight controls are mechanical hydraulically powered. It is powered by two Klimov /Sarkisov RD-33 turbofan engines 18,300 pounds st with afterburners. Fuel capacity is 1153 gallons with attachment for 396 gallon external fuel tank under fuselage. Some aircraft are equipped to carry 304-gallon external fuel tanks under each wing.

It has single point refueling in the port wheel well and overwing receptacles for manual refueling. It also has a turbo-shaft 98 shp APU for engine starting, located above rear on port side of fuselage.

MiG-29 specifications are as follows: Wings are swept back with a span of 37 feet, 3 inches and overall length including the nose probe is 56 feet 10 inches. Without the nose probe it is 53 feet 5 inches. Operating weight empty is 24,030 pounds, maximum fuel load is 10,230 pounds, maximum weapon load is 6615 pounds and normal takeoff weight is 33,600 pounds. Maximum takeoff weight is 40,785 pounds. Maximum level speed at height is Mach 2.3 or 1320 knots or 1520 mph. At sea level, it is Mach number 1.06, 700 knots or 805 mph. Takeoff speed is 119 knots or 137 mph. Landing speed is 127 knots or 162 mph. Takeoff run is 8290 feet and landing run is 1970 feet with brakes and chute. Range is 932 miles with maximum internal fuel and 1305 miles with underbelly tank. G limits are +7 above Mach 0.85 and +9 below Mach 0.85 It can be assumed that this is a supersonic airplane, but some features are not available.

MiG-31, NATO designation Foxhound, is a two-seat twin engine strategic all altitude, all weather interceptor. The MiG-31M is designated as Foxhound B, is an improved interceptor and can be guided automatically, and to engage targets under ground control. The landing gear is retractable tricycle type with dual wheels on main gear retracting forward into air intake trunk and facilitates operation from un-prepared ground and gravel. The nose wheel is a twin unit with mudguard and is rearward retracting. It is powered by two Aviadvigatel D-30F6 turbine engines, each with 34,170 pounds st with afterburners. Internal fuel capacity is about 5350 gallons and provisions for two under wing tanks, each 660 gallons. It has a semi-retractable flight refueling probe on port side of front fuselage. Armament consists of infrared and radar countermeasures and improved missiles and guns. Empty weight is 48,115 pounds, internal fuel is 36,045 pounds and maximum takeoff weight is 90,390 pounds with internal fuel and 1101,850 pounds with maximum fuel and two underwing tanks. Maximum permitted speed is Mach 2.83, maximum level speed at 57,400 feet altitude is 1620 knots or 1865 mph. At sea Level, it is Mach 0.85. Service ceiling is 67,600 feet altitude and maximum takeoff run at maximum takeoff weight is 3940 pounds. Landing run is 2625 feet. Ferry range with maximum fuel and no missiles is 2050 miles. Maximum endurance with external tanks is 3 hours and 36 minutes, and refueled in flight is 6 to 7 hours. Maximum g limit at supersonic speed is +5.

Mikoyan 1-42 is a new generation of the Russian Su-27/35 and MiG-29. It is a single seat multi-role combat aircraft. It is planned to be powered by two Lyulka AL-41F turbofans equipped with afterburners. The use of radar absorbent materials, rather than radar absorbent structures is expected. It is assumed that emphasis on multi-role capability will be made with internal stowage of some air-to-air missiles and a heavy reliance on countermeasures. The wingspan will be similar to the Su-27. It will also include the newly developed phased array fire control radar. Not much else is known, but this is expected to be a highly advanced military fighter and attack airplane.

Mikoyan MiG –AT is a two seat jet trainer and light attack, low wing monoplane with wing root leading edges swept forward and engine intakes over the wings; non-swept tail plane, ventral fin; and tail cone compromised two front-hinged root type airbrakes. It is designed for maneuverability comparable with front line combat aircraft. It is intended to have a service life of 10,000 hours with up to 25,000 landings. Maximum angle of attack is aimed at 25 degrees and it can sustain a 4g turn at Mach 0.7 Onboard simulation and system failures are to be via HUD. It has fly-by-wire controls. Production aircraft are expected to have composite wings made in South Korea. It has a retractable tricycle type landing gear with a single wheel on each unit. Main landing gear retracts inward and the nose wheel retracts forward. It has high-efficiency brakes and is capable of operation from unpaved surface runways. The canopy is bird-proof and the crew includes a pilot and one observer/

weapons expert, with seats in tandem zero/zero ejection system. The rear seat is raised to improve occupant's forward view. Power is supplied by two Turbomeca Larzac 04-R20 turbofans mounted above wing roots. Each engine produces 3175 pounds st The fuel supply is contained in two fuselage tanks with a total capacity of 1874 pounds and one in wing with a capacity of 441 pounds. The tanks are pressurized with engine bleed air to insure high altitude supply. The wingspan is 32 feet 9 inches, and overall length is 37 feet 1/4 inches. Normal takeoff weight is 10,163 pounds and maximum takeoff weight is 12,037 pounds. Maximum level speed is 460 knots or 528 mph, landing speed is 92 knots or 106 mph and service ceiling is 49,200 feet. Ferry range is 1865 miles and g limits are +8/-3. The latest is the Mikoyan MiG-110, which is a projected two turboprop multi-purpose transport. This move by MiG is characteristic of many warplane manufacturers to diversify. The cabin can be equipped for passenger, cargo/passenger or cargo applications It is projected to be a high wing monoplane with a anhedral section between twin booms carrying engines at front and twin fins at the rear. It will have a retractable tricycle landing gear with twin wheels on each unit. The main wheels retract into booms rear of engines. The fuselage pod is slung from center section with beaver-tail rear loading ramp. It is powered by two 2465 shp Klimov TV7-117V turboprops with six-blade propellers. The accommodations are for 35 passengers, or 15 passengers and 7715 pounds of freight, or 11,025 pounds of freight. The wingspan is 72 feet 7 inches, overall length is 60 feet inches and maximum takeoff weight is 33,730 pounds. Estimated nominal cruising speed at 36,000 feet is 270 knots or 310 mph. Range with maximum fuel is projected to be 2515miles.

MIL [RUSSIA]

This OKB was formed in 1947 by Mikhail Leontyevich Mil who was involved with the Russian gyroplane and helicopter development from 1929 until his death in 1970 at the age of 60.

The original Mi-1 was first flown in 1948 and began service in 1951 and was the first in a series helicopter production in the former USSR. More than 25,000 helicopters of Mil design have been built representing 95 percent in CIS. Mil design and production were integrated into a group including Mil Moscow, Kazan and Rostov plants, a helicopter operating company, financial and insurance interests. Associates include Ulan-Ude production center, Arsenyev and Viatka factories. The MIL Mi-2 (V2) was built in Poland and described under the Polish section of PZL Swidnik.

Mil Mi-6 and Mi-22, NATO named Hook, were developed by a joint civil and military requirement in 1954. The first prototype flew in 1957 and was the largest helicopter in the World of that time. More than 800 were built for civil and military use ending in 1981. Later developments include Mi-10 and Mi-10K flying crane, Mi-6 dynamic components were duplicated on the V-12 (Mi-12) of 1967, which remains the largest helicopter flown so far. TheMi-6, Hook A, basic transport description applies to this version. This series include the Mi-6VKP, NATO Hook-B Command Support Helicopter and the Mi-22 which was the Hook-C. Mi-22 is the command and support version with additional antennae for the CIS ground forces to haul guns, armor, vehicles, supplies, freight and troops in combat areas and in combat support roles. The exports include Algeria, Iraq, Peru, and Viet Nam. It has two shoulder wings offload the rotor by providing about 20 percent of total lift in cruising flight and removed when the helicopter is used as a flying crane. The power plant is two 5425shp Aviadvigatel/Solviev D-25V (TV-2BM) turbo-shaft engines mounted side by side above the cabin forward of main rotor shaft. It has eleven internal fuel tanks with a capacity of 13,922 pounds, and two external tanks on each side of the cabin with a total capacity of 7695 pounds. There are also ferry tanks inside the cabin with a capacity of 7695 pounds. The crew consists of five; two pilots, navigator, flight engineer and radio operator. It is easily converted to accommodate 65 to 90 passengers with baggage or cargo in the aisles. The normal military seating is 70 combat troops, as an ambulance 41 stretcher cases and two medical attendants on tip-up seats that can be carried. This ambulance version also has provisions for oxygen. The cabin floor is stressed for 410 pounds per square foot and provisions are available for tie-down rings. Rear clamshell ramp and ramp are hydraulically operated. It is equipped with an electric winch with a capacity of 1765 pounds and a pulley block system. It has three doors that can be jettisoned fore and aft of main landing gear on port side and aft of landing gear on starboard side. The rotor is five bladed and the diameter is 114 feet 10 inches, tail rotor diameter is 20 feet 8 inches and the length with rotors turning is 136 feet 1 ½ inches. Overall height is 32 feet 4 inches. Empty weight is 60,055 pounds, maximum internal load is 26,450 pounds, maximum sling load is 17,637 pounds, fuel load is 13,922 pounds internally and with external tanks it is 21,617 pounds. Maximum takeoff weight is 84,657 pounds for sling loads below 3280 feet. Normal takeoff weight is 89,285 pounds and maximum takeoff weight for VTO is 93,700 pounds. Maximum level speed is 162 knots or 186 mph, maximum cruising speed is 135 knots or 155 mph. Service ceiling is 14,750 pounds. range is 334 nautical miles or 385 miles with 17,637 pounds payload. With external tanks and 9920 pounds payload it is 540 nautical miles or 621 miles. Maximum ferry range with tanks in cabin is 781 nautical miles or 900 miles.

MiL-8, reporting name Hip, is twin turbine multi purpose helicopter. The versions are:Mi-8 is a civil passenger helicopter seating 28 to 32 persons in main cabin with large square windows; Mi-8T Hip C is a civil version normal payload of internal or external freight, but with 24 tip-up seats along cabin sidewalls being optional; Mi-8TG has modified TV2-117TG engines permitting operation on LP gas and kerosene. LPG is contained in large tanks on each side of the cabin under low pressure. Engine operation requires engines to switch to kerosene for takeoff and landing. Weights were not changed; Mi −8 Salon is a deluxe version of the Mi-8with all the amities of executive travel. The military version is a standard assault transport of CIS army support forces with a twin rack for stores on each side to carry rockets in four-packs or other weapons. The military versions, Mi-8MT and Mi-8MTV can be identified by the port side tail rotor. Other military designations were Hip-D, Hip-E, Hip-F, Hip-G, Hip-H, and Hip-K [Mi-8PP.

Customers include CIS ground forces who have about 2400 Mi-8 and Mi-17. At least 40 other Air Forces and civil operators worldwide operate them. The HIP series landing gear is non-retractable, tricycle steerable twin nose wheel locked in flight. The main gear has a single wheel on each unit. There are pneumatic brakes on main landing gear and the pneumatic system can also recharge tires in the field using air stored in the main landing gear struts. is powered by two 1677 shp Klimov TV2-117A turbo-shaft engines. The Mi-8MT has two 1923shp TV3-117MT engines. The main rotor speed is governed automatically with manual override. It has a single flexible internal fuel tank holding 117.5 gallons; two external tanks on each side of cabin holding 197 gallons in port tank and 179.5 gallons in the starboard tank. Total fuel capacity is 494 gallons. There are provisions for one or two ferry tanks in the cabin, raising total fuel capacity to 977 gallons. A fairing over the starboard tank houses optional cabin air conditioning equipment at front. Engine cowling side panels form maintenance platforms when open. Total oil capacity is 132 pounds.

The five-blade Main rotor diameter is 69 feet 10 1/4 inches, tail rotor is 12 feet 9 7/8 inches in diameter and overall length with rotors turning is 82 feet 9 inches. Width of fuselage is 8 feet 2 ½ inches. The passenger cabin is 20 feet 10 inches, height is 7 feet 8 inches. Cargo hold for the freighter version is 17 feet 6 inches long, 7 feet 8 inches wide and 5 feet 10 3/4 inches high. Empty weight for the passenger version is 14,990 pounds. For the civil cargo version the weight is 14,603 pounds. The typical military version weighs 16,007 pounds. Maximum payload internally is 8820 pounds and externally 6614 pounds. Fuel with standard tanks is 3197 pounds and with two auxiliary tanks 6327 pounds. Normal takeoff weight is 24,470 pounds, with 28 passengers each with 33 pounds of baggage is 25,508 pounds, and with 5510 pounds of slung cargo, 25,195 pounds. Maximum takeoff weight for VTO is 26,455 pounds. Maximum level speed at 3280 feet altitude is 140 knots or 161 mph. Maximum at sea level is 142-155 knots. With 5510 pounds of slung cargo, 79 knots or 112 mph. Maximum cruising speed is 119 knots with normal AUW, and with maximum AUW 97, knots or 112 mph. Service ceiling is 13,125-14765 feet. Range of cargo version, standard fuel and 5 percent reserves, 242 nautical miles or 280 miles. With 24 passengers, 3280 feet altitude and 20 minutes fuel reserve, 270 nautical miles or 311 miles. The cargo version with auxiliary fuel and 5 percent reserve, the range is 518 nautical miles or 596 miles.

The military versions have mission equipment compatible with their mission. Hip C is a standard assault transport for CIS army support forces. It has twin-rack stores on each side to carry 128 57mm rockets in four packs, or other weapons. More than 1500 are in service with CIS and some have been upgraded to Mi-17 as MI-8MT and Mi-8MTV, identified by their port side rotor. The HIP D is an airborne communications version but with rectangular section canisters on outer stores racks, and antennae above and below the forward part of the tail boom. The HIP-E has a flexible mounted 12.7 machine-gun in the nose, triple stores rack on each side to carry 192 rockets in six-packs plus four M17P Scorpion (AT-2 Swatter) anti-tank missiles on rails above the racks. About 250 in CIS ground forces, some upgraded to Mi-17 standard with the tail rotor on the port side. The HIP F is an export version of HIP-E. The missiles are changed to six M14 (NATO AT-3 Saggers) with manual command and line of sight HIP-G and H are described as MI-9 and MI-17 respectively. The HIP J is an ECM version with additional small boxes on each side of the fuselage, fore and aft of main landing gear legs. The HIP-K (Mi-8PP) is an ECM communications jammer. Rectangular container and array of six cruciform dipole antennae are on each side of the cabin. No Doppler box is under tail boom, heat exchangers are under front fuselage. Some were upgraded to Mi-17 standard.

MIL Mi-17 and Mi-171 are twin turbine multi purpose helicopters. Mi-17 is a mid-life update of Mi-8, and includes more powerful turbo-shaft engines in CIS as Mi-8MT and Mi-8 MTV. The engines are TV3-117 VM, 2070shp turbo-shafts resulting in improved rates of climb and hover ceilings. Mi-17VA is a version for the Ministry of Health of the former USSR as a flying hospital to highest standards possible. Developed in Hungary, has provisions for three stretchers, operating table, extensive surgical and medical equipment and accommodations for a doctor/ surgeon and three nursing attendants. Mi-172 is the Mi-17 and Mi-17V has the upgraded engines providing cruising speed of 118 knots or 135 mph and service ceiling of 19,685 feet, with air conditioning, and heating, main and tail rotor deicing, canopy demisting and heating of engine air intakes. Options include flotation gear, Doppler, weather radar, DME, GPS, VOR and LS transponder and VIP interiors accommodating seven, nine and eleven passengers. The following are the characteristics of Mi-17 and –171; Customers include the CIS armed forces, Angola, Cuba, Czech Republic, Hungary, India, North Korea, Nicaragua, Papua New Guinea, Peru, Poland, and Slovaki. More than 810 units were exported by Aviaexport and up to 60 are operational by Sky Link Aviation and Transportation Services of Canada. The TV3-117MT engines are designed so that if one engine would stop, the other would automatically increase power contingency rating to 2195 shp enabling flight to continue. The APU is used for engine starting and deflectors on the engine air intakes prevent ingestion of sand, dust and other foreign objects. Empty weight is 15,653 pounds, internal fuel is 4469 pounds, internal fuel plus one auxiliary tank is 6034 pounds and internal fuel plus two auxiliary tanks, 7600 pounds. Normal takeoff weight is 24,470 pounds to 28,660. Maximum level speed is 124 to 135 knots or 143 to 155 mph. Maximum cruising speed is 113 to 129 knots or 130 to 149 mph. Service ceiling is 11,800 feet to 18,700 feet depending on the version of Mi-17/171. Range at 1640 feet, maximum AUW and 5 percent reserves is 251 nautical miles or 289 miles and with normal AUW, 267 nautical miles or 307 miles. Range at 1640 feet, maximum AUW, 30- minute reserve fuel, internal tanks 307 nautical miles or 354 miles; with internal fuel plus one auxiliary tank 440 nautical miles or 506 miles and internal plus two auxiliary tanks, 575 nautical miles or 661 miles.

MIL Mi-24 (NATO name HIND) is a twin turbine gunship with transport alternative. It is the first Russian helicopter to be designed especially for military combat. This is probably the result of development of the U.S. AH-1 and AH-64 gunships, and the trend of world aviation development. Mi-24A, B, and C (HIND A, Band C) were early versions with pilot and copilot/ gunner seated in tandem under large area continuous glazing, large flight deck and not as heavily armed as the Mi-24D (HIND D). The Mi-24D updates included upgraded TV3-117 engines and port side tail rotor. Entire front fuselage is redesigned above the floor forward of engine air intakes. Heavily armored separate cockpits for weapon operator and pilot in tandem. A flight mechanic is optional in main cabin, but transport capability is retained. It has an under nose JakB-12.7 four-barrel machine gun in turret slaved to the adjacent electro-optical sighting pod for air-to-air and air-to-surface use. Also it has a Phalanx anti-tank missile system, and the nose wheel leg is extended to increase ground clearance of sensor pods and nose wheels are partially exposed when retracted.

Mi-24DU is a training version with no gun turret in the nose. Mi-24V is the HIND E, is the same as Mi-24D but with modified wingtip launchers and four under-wing pylons. Weapons include up to 129 M114 (NATO AT-6 Spiral) radio guided tube tube-launched anti-tank missiles in pairs in the attack missile system, enlarged under nose missile guidance pod on port side with fixed searchlights rearward, K-60 (NATO AA-8 Aphid) air-to-air missiles are optional on under wing pylons and Pilot HUD replaces the former reflector gun sight.

Mi-24VP is another variant of Mi-24 with twin barrel 23mm GSh-23 gun with 450 rounds instead of four barrel 12.7mm gun in the nose. The Mi-24P (Hind F) is the same as Mi-24V but the nose gun turret is replaced with a GSh-30-2 twin-barrel 30 mm gun with 750 rounds in a semi-cylindrical pack on the starboard side of the nose. The bottom of the nose is smoothly faired above and forward of sensors. The "P" designation is referring to the Pushka cannon. Mi-24R has no electro-optical and missile guidance pods. Replacing the wingtip weapon mounts is a clutching hand mechanism on lengthened pylons to obtain six soil samples per sortie for NBC warfare analysis. It has a lozenge shape housing with exhaust pipe of air filtering system under port side of the cabin, small rearward fairing marker flare pack on the tail skid and deployed individually throughout CIS ground forces. Mi-24K is the same as Mi-24 except it has a camera in the cabin with lens on the starboard side for reconnaissance and artillery spotting. It has no target designator pod under the nose, and it has an upward hinging cover for the IR sensor.

Mi-25 is an export version, including for Afghanistan, Cuba and India. Mi-35 is an export of Mi-24V. Mi-35P is an export version of Mi-24P. The Mi-24/35 is designed for the latest air mobility requirements of the Russian Army. The features include Mi-28 main and tail rotors and transmissions, a 23mm nose gun same as MI-24VP, and 9K114-9 Attack laser beam riding developments of the tube launched AT-6 Spiral anti-tank missile. More than 2300 were produced at Arsenyev and Rostov. Other customers for this series are Afghanistan, Algeria, Angola, Bulgaria, Cuba, Czech Republic, former East Germany, Hungary, India, Iraq, North Korea, Libya, Mozambique, Nicaragua, Peru, Poland, Vietnam and Yemen. Specifications for the latest Mi-24/35 are the same but constant upgrade makes present status unavailable. The rotor system is a constant chord, five blade main rotor. Each blade has aluminum alloy spar, skin, and honeycomb core. The spars are nitrogen pressurized for crack detection. They have a three blade tail rotor, main rotor brake, and 5mm hardened steel side armor on the front fuselage. The landing gear is tricycle type, retracting rearward twin nose wheel unit. It has Single main units with pneumatic shock absorbers and low-pressure tires. Main units retract rearward and inward into aft of the fuselage pod, turning through 90 degrees to stow almost vertically, disc wise to longitudinal axis of the fuselage under blister fairings. A tubular tripod skid assembly with shock strut protects the tail rotor in tail-down takeoff or landing. The main cabin can accommodate eight persons on folding seats or four stretchers. The cabin and cockpit are heated and ventilated. They have a dual electrical system with three generators providing 36, 115, and 208 volt AC at 400hz and 27 volt DC. There is an electro-thermal de-icing system for main and tail rotor blades.

The APU is mounted transversely inside fairing aft of the rotor head. The main rotor diameter is 56 feet 9 inches and tail rotor diameter is 12 feet 10 inches. Overall length is 70 feet ½ inch with rotors turning, and 57 feet excluding rotors and gun. The main cabin is 9 feet 3 inches long, 4 feet 9 ½ inches wide and 3 feet 11 1/4 inches high. Weight empty is 18,078 pounds, and maximum internal stores is 5290 pounds and normal takeoff weight is 26,455 pounds. Maximum speed is 180 knots level or 208 mph and cruising speed is 145 knots or 168 mph. Economical cruise speed is 117 knots or 135 mph. Service ceiling is 14,750 feet, Hover out-of-ground effect is 4820 feet. Combat radius is 86 nautical miles or 99 miles with maximum military load and with two external fuel tanks, 121 nautical miles or 139 miles and Mig MIL-25 (NATO name HIND-D) is an export version of the Mi-24D for Angola, India and Peru. It can be assumed that variations in equipment standards are likely.MIL Mi-26 is classified as a twin-turbine multi-purpose heavy lift helicopter. Significant models include a passenger version which carries 63 passengers four abreast in airline type seating with center aisle, toilet galley, and cloakroom aft of the flight deck. Another is a medical evacuation version with life support section for four casualties and two medics, surgical section for two casualties and two medics, ambulance section for five stretcher patients, three seated casualties and two attendants, laboratory, and amenities section with toilet, food storage and recreation unit. Another version serves as a tanker with internal fuel capacity of 3710 gallons of fuel and provisions for four auxiliary tanks, as well as lubricants dispensed through four hoses for ground vehicles. The power plant is two 10,000 shp ZMBK Progress D-136 free turbine turbo-shafts side-by-side above the cabin and forward of the main rotor driveshaft. The landing gear is a tricycle non-retractable tail wheel type with a single wheel on each unit. It has two Klimov 2070shp turbo-shaft engines in pod above each wing root. An APU in the rear of main pylon supplies compressed air for engine starting and drive a small turbine for preflight ground checks. There are deflectors for dust and foreign objects forward of air intakes which are deiced by engine bleed air.

The main rotor is 56 feet 5 inches in diameter, the tail rotor diameter is 12 feet 7 inches and overall length excluding rotors is 55 feet 9 inches. Empty weight is 17,846 pounds, internal fuel capacity is 2947 pounds and with added tanks, 3928 pounds. Normal takeoff weight is 22,928 pounds, maximum takeoff weight is 25,705 pounds, maximum level speed is 162 knots or 186 mph and maximum cruising speed is 145 knots or 167 mph.. Service ceiling is 19,025 feet and radius of action with standard fuel, 10 minutes loiter and 5 percent reserves is 108 nautical miles or 124 miles. Range with maximum standard fuel is 248 nautical miles or 285 miles. Ferry range with 5 percent reserve is 593 nautical miles or 683 miles. Endurance with maximum fuel is 2 hours and g limits are +3 and –0.5.

MIL Mi-34, NATO name Hermit, is a lightweight two/four seat multipurpose helicopter. It is skid equipped with a tailskid to protect the tail rotor and is used as general purpose helicopter in such roles as private, executive, training, police, patrol, ambulance, international competition and observation. It is capable of performing loop and roll acrobatic maneuvers. The fuel system is designed for inverted flight. The controls are manual with no hydraulic boost. It is powered by one 325hp VMKB M-14V-29 radial engine with nine cylinders. The main rotor is four-blade and the tail rotor is two-blade. Main rotor is 32 feet 9 inches in diameter and the tail rotor diameter is 4 feet 10 inches and the fuselage is 28 feet 8 ½ inches long, 4 feet 7 inches wide. The cabin is heated, and main and tail rotor can be de-iced. Fuel weight is 282 pounds. Takeoff weight is 2425 pounds for aerobatic, normal is 2822 pounds and maximum is 2976 pounds. Maximum level speed is 118 knots or 136 mph; maximum cruising speed is 97 knots or 112 mph, normal cruising speed is 86 knots or 99 mph and service ceiling is 14,765 feet. Range at takeoff weight with 2976 pounds and 363 pounds payload is 97 nautical miles or 112 miles. With 320 pounds payload it is 194 nautical miles or 224 miles.

MIL Mi-34 VAZ is a lightweight two/four seat twin-engine helicopter. It is basically the same as Mi-34 but with a completely new rotor head making the main rotor tip speed 672 feet per second. It is equipped with two VAZ-430 twin-chamber rotary engines, each producing 227hp for takeoff and 266hp for contingency. It is equipped with optional stretcher for EMS duties. These changed the main rotor diameter to 37 feet 5 inches, fuselage length to 31 feet 2 inches. Maximum takeoff weight is 4320 pounds, maximum level speed is 113 knots or 130 mph and nominal cruising aped is 97 knots or 112 mph. Range with maximum payload and 30 minute reserve is 43 nautical miles or 50 miles. With 750 pounds of payload and 30 minutes reserve, it is 323 nautical miles or 372 miles. Mi-35 is the export version of Mi-24V and Mi-35P is the export version of the Mi-24P.

Mi-38 is a twin turbine multi-role medium range helicopter. It has a conventional pod and boom configuration with the power plant above the cabin. The six-blade main rotor has significant non-linear twist and swept tips. It also has two independent two-blade tail rotors set as narrow X on the same shaft. It is equipped for cargo airdrop and cargo sling attachment. The controls are fly-by-wire, which means no mechanical controls. Main and tail rotor blades are made of composites and the fuselage is mostly made of composites.. It has a retractable tricycle landing gear, single wheels on the main units and twin nose wheels. Optional pontoons are available for emergency use in over water missions. The two engines are Klimov TVA-3000 with 2465shp for takeoff. Single engine rating is rated at 3500 shp and the transmissions are rated for the same. The power plant is located above the cabin and to the rear of the main reduction gear. Bag fuel tanks are beneath the floor of the main cabin. Provisions are included for external auxiliary fuel tanks. Liquid Petroleum Gas (LPG) is planned as alternative to aviation kerosene. It accommodates a crew of two on the flight deck, which is separated from the main by a compartment for the majority of avionics. It has lightweight seats for 30 passengers as alternative to unobstructed hold for 11,020 pounds of freight. Air survey and ambulance versions are planned. There are provisions for a hoist over the port side door, powered and overhead rails in the cabin. It has deicing for main and tail rotor blades and air conditioning by compressed bleed air or the APU on the ground.

Main rotor is 69 feet 2 inches in diameter, and the tail rotor diameter is 12 feet 7 inches. Overall length excluding rotors is 64 feet 7 ½ inches. Maximum internal width is 7 feet 9inches, and height in the center is 5 feet 11 inches. Maximum takeoff weight is 34,170 pounds, normal takeoff weight is 31,305 pounds and maximum payload internal or external is between 11,020 and 13,225 pounds. Maximum level speed is 148 knots or 171 mph and cruising speed is 135 knots or 155 mph. Service ceiling is 21,325 feet hover out-of-ground-effect is 8200 feet. Range depending on payload and fuel on board is from 175 nautical miles or 292 miles, to 700 nautical miles or 808 miles.

Mi-40 is a twin turbine infantry combat helicopter based on the concepts of Mi-24 and Mi-28. The landing gear is retractable nose wheel type, with twin wheel nose unit and a single wheel on each main unit. The power plant is the same as for the Mi-28N. It accommodates a pilot and navigator/weapon systems operator side by side and up to seven combat equipped troops in the main cabin. There is a compartment in the rear cabin for a gunner. Armament consists of an under-nose turret for 23mm multi-barrel gun, up to eight air-to-surface or air-to-air missiles, gun and seven troops. There are provisions for flexibly mounted 12.7mm gun at the rear of the cabin pod. The main rotor is 56 feet 5 inches in diameter and tail rotor diameter is 12 feet 7 inches. Empty weight is 16,920 pounds, maximum payload is 3968 pounds, maximum internal fuel is 2580 pounds and maximum takeoff weight is 26,235 pounds. Maximum level speed is 167 knots or 192 mph. Nominal cruising speed is 140 knots or 161 mph. Service ceiling is 18,200 feet, Hover OGE is 10,825 feet, and maximum range is 215 nautical miles or 248 miles.

Mi-46T is a twin-engine turbo-shaft passenger or freight transport helicopter. It is a replacement for the Mi-6. It has a seven-blade main rotor, five-blade tail rotor and retractable tricycle landing gear with twin wheels on each unit and with rear loading ramp. The power plant is two new type Aviadvigatel turbo-shaft engines, each generating 7495shp. It is normally equipped to carry freight. Main rotor diameter is 90 feet 6 inches in diameter and length overall excluding rotors is 86 feet 3 ½ inches. Empty weight is 35,715 pounds, maximum payload is 26,455 pounds and maximum takeoff weight is 66,137 pounds. Nominal cruising speed is 145 knots or 167 mph. Hover ceiling at maximum weight is 7545 feet and range with 22,045 pounds payload is 248 miles.

Mi-46K is a twin-turbine flying crane helicopter. It is a replacement for the Mi-10K and is a typical flying crane. Flight deck, power plant and all dynamic components are the same as Mi-46T. It has short stub wings and long tripod braced main wheel units and a glazed gondola for second pilot facing rearward behind the front fuselage pod to control the helicopter during loading and unloading. It also has a sling cable directly under the main rotor drive shaft for payload. The power plant is two turbo-shaft engines, upgraded to produce 8,000 shp. Empty weight is 43,430 pounds and maximum takeoff weight is 80,467 pounds. Hovering ceiling and range are similar to the Mi-46T.

Mi-52 is a three seat light helicopter. It is the smallest helicopter developed by MiL. It has a four blade main rotor and two-blade tail rotor. The landing gear is non-retractable tricycle type, with a single wheel on each unit with cantilevered spring legs and fairing over each heel. It is powered with one rotary Wankel-type engine. It has an enclosed cabin for the pilot in front and two passengers on the rear bench seat. Also there are two doors on each side. Diameter of main rotor is 32 feet 9 inches and overall length, excluding rotors, is 28 feet 7 inches. Maximum payload is 551 pounds and maximum takeoff weight is 2535 pounds. Nominal cruising speed is 86 to 91 knots, or 100 to 105 mph. Hover OGE is 5250 feet and maximum range with standard fuel is 215 nautical miles or 248 miles.

Mi-54 is a twin-turbine commercial utility helicopter. It includes four blade composite main and tail rotor blades It is intended for passenger/cargo, oil rig support, ambulance and executive duties. It has a non-retractable tricycle type landing gear with a single wheel on each main unit, the nose wheels are twin and it has a tailskid. It is powered by two Saturn/Lyulka AL-32 turbo-shaft 770shp engines are mounted side by side above the cabin.. Main rotor diameter is 44 feet 3 inches, The tail rotor is 8 feet 6 ½ inches in diameter and overall length excluding rotors is 43 feet 3 inches. Payload is 2205 to 2865 pounds and maximum takeoff weight is 8820 pounds. Maximum level speed is 151 knots or 174 mph, normal cruising speed is 140 knots or 161 mph and hover OGE is 6560 feet. Range with maximum internal fuel is 431 nautical miles or 497 miles and with 2645 pounds payload is 242 nautical miles or 280 miles.

MOLNIYA [RUSSIA]

Molniya Scientific and Industrial Enterprise is located in Moscow. To compensate for reduced funding in the Russian space program, they have gone to a series of civil aircraft. The first is the Molniya-1 which had six-seats. Following this, Molniya-100 was a 15 seat configuration,

Molniya-300 is a twin-engine business jet, Molniya-400 is a twin-engine cargo jet, and the six engine Molniya-1000 Hercules freighter would be the world's heaviest airplane. The Hercules has a maximum takeoff weight of 1,984,125 pounds and payload of up to 992,060 pounds suspended under its inverted gull-wing center section. Payloads would include a shuttle orbiter or large rocket stages.

Molniya-1 has a power plant prototype VMKB M-14PM-1 air-cooled radial piston engine driving a three-blade pusher propeller. An alternative is the Teledyne Continental TSIOL-550 horizontally opposed engine. It was initially not supercharged but supercharged later. It accommodates six persons in pairs. The seats are readily removed to convert to cargo. The wing span is 27 feet 10 inches and overall length is 25 feet 9 ½ inches. Maximum payload is 1115 pounds, maximum fuel is 485 pounds and maximum takeoff weight is 3835 pounds. With the Continental engine, maximum level speed is 221 knots or 255 mph. Nominal cruising is 145 to 191 knots or 168 to 220 mph, landing speed is 68 knots or 78 mph. Takeoff run is 1150 feet, and range with maximum payload at 145 knots ore 168 mph is 270 nautical miles or 310 miles and with maximum fuel is 648 nautical miles or 745 miles.

Moliya-100 is a twin-engine multi-purpose light transport. It has a retractable tricycle type landing gear with a single wheel on each unit. The nose gear retracts forward. The power plant has two VMK M-14PM NTK air-cooled radial piston engines, with superchargers and a direct injection fuel system, or two Teledyne Continental GTSIO-520 flat-six engines and later option includes turboprop engines The accommodations include two crew side by side with 15 seats in the cabin. The ambulance configuration includes space for four litter patients, two medical attendants and medical equipment. It is basically IFR, but GPS and other systems are optional. Wingspan is 46 feet 11 inches and overall length is 40 feet 11 ½ inches. Maximum payload is 3307 pounds, maximum fuel is 2095 pounds and maximum takeoff weight is 9480 pounds. Maximum level speed is 216 knots or 248 mph and nominal cruising speed at 9840 feet is 175-205 knots or 202-236 mph. Landing speed is 70 knots or 81 mph, takeoff run is 1410 feet, landing run is 1150 feet and range with maximum payload is 323 nautical miles or 372 miles and with maximum fuel is 1080 nautical miles or 1242 miles.

MOONEY AIRCRAFT CORP. [USA]

The original Mooney Company was formed in Wichita, Kansas in 1948. They produced a single-seat M-18 Mite until 1952. In 1985, Alexandere Couvelaire, president of Euraliar of Paris, France and Michel Seydoux, president of MSC jointly acquired Mooney. Mooney produced the Mooney 205 MSE, a four seat touring aircraft and the 205 MSE, 201 ATS. Later the newly formed. Mooney produced the Mooney 20R Ovation, also a four seat touring aircraft. They produced the Mooney TLS which was turbocharged. It was a stretched version of the M20M. It had one 270 horsepower Textron Lycoming TIO-540-AFIA flat-six engine driving a McCauley three-blade propeller, oxygen system and upgraded avionics. They are still popular with small operators and private pilots.

MYASISHCHEV [RUSSIA]

Myasishchev is located, as most Russian are aircraft companies are, in the Moscow region. It was founded in 1851 and in the 1950's it developed and built the M-4 and 3M (NATO Bison) subsonic strategic bombers and M-50 (NATO Bounder) supersonic strategic missile carrier. In the 1970's and 1980's the bureau was engaged in development of multi-purpose sub-sonic high altitude aircraft. They modified two 3M into VM-T Atlant aircraft to transport sections of the Energia rocket launch vehicle and airframe of the Buran space shuttle orbiter. In 1981 the design bureau was named after Professor V.M. Myasishchev who died in 1978.

VM-T Atlant has four RKBM/Koliesov VD-7MD turbo-jet engines each producing 23,700 pounds st. It has the capability to transport above-fuselage cargoes of four types with maximum weight of 110,230 pounds and maximum diameter of 26 feet 3 inches. The wing span is 174 feet 5 inches, overall length is 167 feet 11 3/4 inches and height is 34 feet 9 ½ inches. Maximum takeoff weight is 304,233 pounds, maximum cruising speed at 19,700 to 26,250 feet altitude is 270-323 knots or 310-372 mph. Runway length needed is 11,500 feet and range with maximum payload is 810 nautical miles or 932 miles.

M-17 and M-55, NATO name Mystic-A, are single-seat high altitude reconnaissance and research aircraft. M-17 was named Stratosphere. It is equipped with one RD-36-51V turbojet engine rated at 15,430 pounds st in non-afterburning form. The M-55 Geophysica (Mystic A) is basically M-17, but with two PS-30V12 engines side by side at the rear of the fuselage pod and with 11,025 pounds st each. The M-55 can loiter for 4 hours and 22 minutes at 65,000 feet altitude With 3305 pounds of sensors it can loiter for 5 hours at 17,000 feet altitude. It has a retractable tricycle landing gear with twin wheels on each unit. All units retract rearward, main wheels into tail booms. Mystic A wingspan is 133 feet 6 ½ inches and Mystic B is 122 feet 11 inches. Overall length is 69 feet 6 ½ inches for Mystic A and for Mystic B excluding probes is 75 feet inch. Payload for the Mystic B is 3307 pounds, and maximum takeoff weight for Mystic A is 44,000 pounds. Maximum level speed for Mystic B is 377-405 knots or 435-466 mph. Service ceiling for Mystic B is 65,600 feet, landing run is 5745 feet and maximum endurance is 2 hours and 30 minutes for the Mystic A at 55,775 feet altitude and for the Mystic B it is 6 hours and 30 minutes.

M-101 GZEL is a multi-purpose 8 seat turbo-prop aircraft. It is of conventional all-metal low wing monoplane, with swept back vertical tail surfaces. The landing gear is retractable tricycle type with a single wheel on each unit. The nose wheel retracts rearward and main wheels retract inward into wing roots and fuselage. It is designed to operate from unimproved runways. It is powered by one 778 shp Motorlet M-601F turboprop engine driving a AV-510 five blade propeller. Later versions ware planned to use Pratt & Whitney PT-6A turboprops. Accommodations are for one or two pilots and 6 or 7 passengers in a pressurized cabin, It can be rapidly changed from passenger to cargo or ambulance configurations.

MM-1 is a twin turboprop light multi-mission transport. It is designed for a fail-safe more than 30,000 one-hour flights without any major repair. The landing gear is hydraulically retractable tricycle type with a steerable twin nose wheel. Single main wheels retract into fuselage pods and it is equipped with hydraulic brakes. It is powered by two Pratt & Whitney of Canada 1657shp PT6A turboprop engines, with five or six blade propellers, constant speed, fully feathering, reversible pitch of advanced design to minimize noise levels. It is equipped with six integral wing fuel tanks holding 660 gallons. Optional external tanks are available, for either pressure or gravity filling. The airplane is pressurized and air-conditioned. An APU is optional. It has electric de-icing of windscreen and propellers. The pitot is electrically heated and the wind screen is demisted by hot air. Wingspan is 63 feet and overall length is 59 feet 10 ½ inches. Operational weight empty is 10,582 pounds, maximum payload is 5290 pounds, maximum fuel is 4410 pounds and maximum takeoff weight is 20,282 pounds. Maximum level speed is 243 knots or 280 miles, Takeoff run is 2870 feet and landing run is 2035 feet.

M-200 is a two seat military and civil advanced jet trainer. The design life is 15,000 hours of flight. It is powered by two RD-35 turbofans modified Klimov ZMKB Progress DV-2, each producing

4850 pounds st Empty weight is 8390 pounds and maximum takeoff weight is 10,360 pounds. Nominal cruising speed is 378 knots or 435 mph, service ceiling is 42,980 feet, takeoff run is 656 feet, landing run is 1575 feet and range with maximum fuel is 756 nautical miles or 870 miles.

NASA [USA]

The National Aeronautics and Space Administration has headquarters in Washington DC is not primarily a manufacturer of aircraft, they are the only developer of Space Vehicles. In support of the space mission, they operate a fleet of about 112 aircraft, about 34 dedicated for research and 70 for program support. This requires NASA to employ 137 pilots, including 45 astronauts. Most aircraft carry civilian registrations according to their operating bases. They are: Ames Research Center, Moffett Field, California; Dryden Flight Research Facility, Edwards AFB, California; Langley Research Center, Hampton, Virginia; Lewis Research Center, Cleveland, Ohio; Johnson Space Center, Houston, Texas; Wallops Flight Facility, Wallops Island, Virginia and the Kennedy Space Center, Florida.

NORTHOP GRUMMAN CORP. [USA]

Northrop Aircraft, Inc. was organized in 1939 by John K. Northrop to produce military aircraft. Activities extended to missiles, target drones, electronics, space technology, communications, support services, and commercial products. In 1959 their name was changed to Northrop Corporation. They acquired Grumman in May 1994. Grumman Aircraft Engineering Corporation, of Bethpage, Long Island was founded in 1929. In 1969, they formed a holding company for Grumman Aerospace Corporation, Grumman Allied Industries, Inc. and Grumman Data Systems.

Ten operating divisions were created in February 1985

followed by further reorganizations in 1987-1988. A reconsolidation of the four Divisions resulted in reduction to three. They were Aerospace and Electronics, Data Systems and Services and Allied. Their more recent production of the Army OV-1 Mohawk was the most complex aircraft developed and tested to date. The Consolidated plan finalized in January 1994 required Grumman to shed one-third of worldwide capacity and abandon independent design and development of aircraft was to close the Calverton plant and transfer activities to St. Augustine, FL, and nonproduction, including engineering, technical and computer, to Bethpage. At the time of takeover, Grumman had 17,900 employees.

The Bethpage plant was closed for aircraft assembly in November 1992, but remains for component manufacture, final assembly and flight test at Government owned Calverton plant and aircraft upgrades at Melbourne, FL. In mid-1994, The new Northrop Grumman organization was structured into five Divisions: B-2, Military Aircraft, Commercial Aircraft, Electronic and Systems Integration, and Data Systems and Services. The consolidation was to be completed by the end of December 1994. At the finish of the consolidation, the total combined workforce of both was 45,000. Major programs include the B-2 Spirit bomber, which is a swept-wing equipped powered by four General Electric F118-GE-110 Turbo Fan engines. It is made with mostly composites to reduce the radar signature. Each engine produces 19,000 shaft horsepower. The fuel was changed from JP-4 fuel to JP-8 in 1996. It is 69 feet long, 17 feet high and wheel track is 40 feet. Normal takeoff weight is 371,330 pounds, maximum is 376,000 pounds, weapon load is 50,00 pounds and maximum fuel load is between 180,000 and 200,000 pounds. Range is Between 5063 and 7250 miles. With one refueling, the range is increased to 11,508 miles.

Another was the Grumman A-6 and AE-6 Intruder and the F-14 Tomcat. Northrop Grumman is heavily involved in subsystem integration development and production, including

J-Stars. The future includes development of E-8 with four Pratt and Whitney engines with 18,000 pounds of thrust. It weighs 171,000 pounds empty, maximum fuel load is 155,000 pounds and maximum takeoff weight is 336,000 pounds. Service Ceiling is 42,000 feet and Endurance is 11 hours and upgraded to 20 hours with one refueling. It provides for 18 operators. It is interesting that the E 6C used second-hand Boeing 707 airframes that were stored at Davis-Monthan AFB, Arizona.

PIPER AIRCRAFT CORP. [USA]

William T. Piper, known as the Henry Ford of aviation in the United States, was a Pennsylvania oilman who joined the Taylor Aircraft Company Board of Directors in the early 1920's. He had no experience with aviation, but Taylor had developed a small monoplane powered by a 20 HP Brownbach Tiger Kitten engine. It was with this early airplane that gave birth to the Piper Cub. In the early 1930's when Gilbert Taylor left the company, Bill Piper hired a new Chief Engineer named Walter Jamouneau, which under his direction the original Cub was modified to the historic bright yellow J-3 version.

In 1937, a fire in the original Bradford factory devastated the organization and the company was relocated to Lock Haven, Pennsylvania and renamed the Piper Aircraft Company. In the early forties, the Piper J-3, with modifications, became the military L-4. It was used to train military combat pilots. Almost 6,000 of the L-4 aircraft were delivered and four out of five World War II pilots got their start in these very dependable airplanes. In 1947 the company was in serious financial trouble. The banks took temporary control of the company and developed the Vagabond which is acknowledged as what saved Piper. This was the first of the short winged Pipers. In the post-war period, civilian aviation took a rapid growth and advances in technology To keep pace with demand. After building tube frames and fabric-covered aircraft for 17 years, Piper built the first all metal plane, the

Apache in 1954. This was the first twin-engine Piper and the first to be named after an American Indian. With the success of the Apache, there was a bright future for expansion. In 1957, they built a new R&D facility at the old Navy Base in Vero Beach, Florida. This proved to be an excellent site for experimental flight testing. The first accomplishment was the PA-25 Pawnee which was an agricultural spraying aircraft. With the advanced thinking in the Vero Beach facility, they introduced the forerunner of a whole family of innovative and successful aircraft. The Cherokee was FAA certified in 1960, and entered production in January 1961. The PA-28 Cherokee formed the basis more than half of Piper's aircraft in the decades to follow. Subsequent models were, the Warrior, Archer Dakota, Arrow, Seneca, and Saratoga. They were all four-place aircraft.

In 1965 they introduced the PA-32, or Cherokee Six that showed off a new stretched cabin that would accommodate six passengers. It had stowage for luggage in a new forward baggage compartment and a larger cabin interior. There was a rear door for passengers to use. In 1967, the twin-engine lineup was enhanced with the introduction of the PA-31 Navajo. This more powerful cabin class twin was designed to meet the growing needs of business. From this start, this was succeeded with the Navajo Chieftain and the Mojave. These were followed by the twin turboprop powered Cheyenne I, Cheyenne II and the Cheyenne II XL.

In 1971 the PA-34 Seneca was introduced. This six-place light twin was built around the Cherokee Six frame. The Seneca has seen duty around the world as an air taxi, charter aircraft, a dependable trainer for pilots stepping up to larger and more advanced multi-engine classes and was very popular among owners and pilots. Today's Seneca V still fits in each of these categories.

Piper introduced the Cheyenne III in 1979 and later upgraded to the Cheyenne IIIA. These aircraft had seating for 11 and could cruise at 300 knots with a range of 2,000 miles. It was equipped with two turbo-prop 720 HP Pratt And Whitney engines. They targeted the executive market and also found their way into many training programs for pilots by such airlines as Luftansa, Atitalia, Korea, Nippon, Austrian, and the Civil Aviation Administration of China. In 1983 when many manufacturers were pulling back or quitting, Piper introduced a totally new aircraft design. The PA-46 was pressurized, six place single-engine which provided the comforts of many small business jets with a fraction of the price and operating cost. In 1988 an upgraded version, the Malibu Mirage, was introduced with a 350 HP turbo-charged Lycoming engine and propelled it to the forefront of the industry. They have been the baseline of small aircraft for more than 80 years the name Piper is known throughout the world. Their present lineup includes the Meridian, Mirage, Seneca V, Saratoga II TC, and Saratoga II HP, Piper 6XT and Piper 6X, Archer III, Seminole, Arrow and Warrior.

ROBIN [FRANCE]

Avions Pierre Robin formed the company in 1957 and the name changed to the founder. 1n 1969. Mr. Robin left the company in 1990. They have delivered about 4,000 aircraft to date and employ about 160 personnel. In 2004 Alpha Aviation in Hamilton New Zealand purchased all of the jigs, tooling, and intellectual rights to the Robin series 2000 series from Robin. These airplanes are two seat all metal trainers and a fully certified variant known as the Robin 2160. Avion Robins was renamed as APEX and continues production of other units, primarily the DR400 series.

The Robin 200 is a two seat light trainer. It has the conventional cable controls and electrically actuated trailing edge slotted flaps. It is all aluminum alloy stressed skin and ribs. The landing gear is non-retractable tricycle type with a single wheel on each unit, and it has hydraulic disc brakes and a parking brake. Power is derived from one Textron Lycoming O-235-L2A flat-four engine, driving a two blade fixed pitch propeller. Fuel tanks hold 31.7

gallons with optional auxiliary tanks. A jettisonable canopy with anti-glare coating is installed, dual stick controls, dual left hand throttles and dual toe brakes. The cabin is heated and ventilated with windscreen defrosting standard. The electrical system is 12Volt. Wingspan is 27 feet 4 inches, overall length is 21 feet 9 ½ inches and overall height is 6 feet 4 ½ inches. Empty weight is 1157 pounds and maximum takeoff weight is 1719 pounds. Cruising speed at 75 percent power is 121 knots or 140 mph. Stall speed is 50 knots or 57 mph. Range is 566 nautical miles or 652 miles and endurance is 4 hours and 35 minutes.

Robin R-2160 is a two seat aerobatic airplane. It has been sold to customers in Europe, Canada, and USA. The controls are conventional as used in all aircraft of this type. It is a semi-monoc oque type with aluminum alloy spars and skin. The landing gear is non-retractable tricycle with fairing, disc brakes are on the main wheels and nose wheel steering is through the rudder bar. Power is derived from one 160 hp Textron Lycoming O-320-D2A flat-four engine and a two-blade fixed pitch propeller. Fuel capacity in the fuselage is 31.7 gallons with option for 42.3 gallons if needed for utility. It accommodates two, seats side by side and has a 12 volt electrical system, avionics include blind flying instruments, VOR, ADF, ILS, GPS, etc. for customer's choice. Wingspan is 26 feet 4 inches, overall length is 23 feet 4 inches and overall height is 7 feet and propeller diameter is 6 feet 2 inches. Empty weight is 1213 pounds and maximum takeoff weight is 1765 pounds for aerobatic and utility. Never exceed speed is 180 knots or 207 mph, maximum level speed is 138 knots or 159 mph. Cruising speed at 7500 feet and 75 percent power is 130 knots or 150 mph; and at 11,000 feet and 65 percent of power, 126 knots or 245 mph. Stall speed is 52 knots or 60 mph and with flaps down, 47 knots or 54 mph. Service ceiling at 100 feet per minute rate of climb is 15,000 feet and range at 65 percent power is 513 miles or 590 miles. G limits are +6 and −3.

Robin DR-400 Dauphin is a three or four seat light training and touring airplane. The flying controls are slab tail plane with trimmable anti-balance tab, interchangeable ailerons, plain flaps, manually operated air brake under main spar outboard of each main undercarriage leg. The structure is all wood single box spar with the ribs threaded over the box, plywood covered leading edge box, fuselage is covered with plywood, and flaps are all metal and are interchangeable. The landing gear is non-retractable tricycle type with hydraulic brakes. The nose wheel steering is through the rudder. Fairings cover all three legs and wheels, brakes are toe operated and a parking brake is standard. Dual controls are standard, the cabin is heated and ventilated, and the baggage compartment access is from the inside. Also, standard equipment includes 12 volt electrical system, audible stall warning and windscreen de-icing. Radio, blind flying equipment, navigation, landing and anti-collision lights are tailored to customer requirements. It is powered by one 112 hp Textron Lycoming O-235-L2A flat-four engine producing 140 horsepower. A fuel tank in the fuselage contains 26.4 gallons and optionally a 3.2 gallon auxiliary tank. The propeller is a two-blade fixed pitch metal or a wood propeller with the same characteristics. Wingspan is 28 feet 7 inches, overall length is 22 feet 10 inches, overall height is 7 feet 3 inches and propeller diameter is 5 feet 10 inches. Empty weight is 1179 to 1279 pounds, maximum baggage is 88 pounds and maximum takeoff and landing weight is 1984 to 2205 pounds. Never exceed speed is 166 knots or 191 mph, maximum level speed at sea level is 130 to143 knots or 150 to 165 mph, maximum cruising speed is 116 to 117 knots or 133 to 134 mph, and stall speed with flaps down is 45 to 47 knots or 51 to 54 mph. Service ceiling is 12,000 to 14,0000 feet, takeoff run is 771 to 804 feet, takeoff to 50 feet is 1591 to 1755 feet, landing run is 656 to 722 feet, and landing from 50 feet is 1510 to 1542 feet. Range with standard fuel at maximum cruising speed and no reserve fuel is 464 nautical miles or 534 miles.

Robin DR 400/160 Major is a four-seat light airplane. It is similar to the Dauphin DR 400 with the following exceptions. Power is supplied by one 160 hp Textron Lycoming O-320D flat-four engine and a two-blade fixed pitch propeller. Total fuel capacity is 50 gallons contained in a fuselage tank and two tanks in wing root leading edges. Provisions are included for an auxiliary tank raising total capacity to 66 gallons. There is seating for four persons on adjustable seats and rear bench seat with maximum load of 340 pounds. It has a forward sliding transparent canopy. and aft of rear seats where four occupants are carried, is a compartment to stow 88 pounds of baggage. Propeller diameter is 6 feet, empty weight is 1257 pounds, maximum takeoff and landing weight is 2315 pounds. Never exceed speed is 166 knots or 191 mph, maximum level speed at sea level is 146 knots or 168 mph, maximum cruising speed at 75 percent power at 8,000 feet is 132 knots or 252 mph, economical cruise is at 65 percent power at 10,500 feet is 30 knots or 150 mph and stalling speed with flaps up is 56 knots or 64 mph, and with flaps down 50 knots or 58 mph. Takeoff run is 1935 feet and landing run is 817 feet. Range with standard fuel at 65 percent power is 783 nautical miles or 900 miles.

Robin DR 400/180 Regent tug and four or five-seat airplane. It is powered by one O-360-A flat four engine. The propeller diameter is 6 feet 4 inches, weight empty is 1322 pounds and maximum takeoff and landing weight is 2425 pounds. Maximum level speed at sea level is 150 knots or 173 mph, maximum cruising at 75 percent power at 7500 feet is 140 knots or 162 mph, and economical cruise at 60 percent power at 12,000 feet is 132 knots or 252 mph. Stall speed with flaps ups is 57 knots or 65 mph and with flaps down it is 52 knots or 59 mph. Service ceiling is 15,475 feet, takeoff run is 1035 feet and landing run is 817 feet. Range with standard fuel at 65 percent power and no fuel reserve is 783 nautical miles or 900 miles.

Robin DR 400 Remo 180R is a glider tug and four or five seat light airplane It is similar to DR 400/180 except the external baggage door and baggage compartment is covered in transparent plexiglas to improve the rearward view and it has a towing hook under the tail. Power is derived from a 180hp Textron Lycoming

O-360-A flat-four engine driving for towing a two-blade propeller. For other purposes a propeller is provided with a coarser pitch. Empty weight is 1234 pounds, maximum takeoff and landing weight is 2205 pounds and maximum level speed is 146 knots or 168 mph. Cruising speed with 70 percent power at 8,000 feet is 124 knots or 143 mph. Stall speed with flaps down is 47 knots or 54 mph and service ceiling is 20,000 feet. Takeoff run from 50 feet altitude towing a single seat sailplane is 1230 feet and landing from 50 feet is 1542 feet. Range at economical cruise with auxiliary fuel and no reserve fuel is 647 nautical miles or 745 miles.

Robin DR 400/200R Remo 200 is a more powerful glider tug based on Remo 180R with a 200hp Textron Lycoming IO-360 engine and a constant speed propeller. It is fitted with a double silencer and noise level has been less than the certification requirement. Based on tug weight of 1754 and a heavy glider of 1323 pounds and takeoff from a short grass runway, the takeoff run is 804 feet and to 50 feet is 1558 feet.

Robin R3000 series are two to four seat light airplanes. The R3000/100 seats two, has a maximum speed of 124 knots, maximum cruising of 113 knots and normal cruising of 108 knots and service ceiling of 13,000 feet. Range with standard fuel at 75 percent of power and no reserve fuel is 605 nautical miles, and with optional fuel, 767 nautical miles. The landing gear is non-retractable tricycle type, and the nose gear is steered with the rudder pedals and locks immediately after takeoff. R3000/100 seats two, three in the R3000/120 and four in R3000/140 and 3000/160. Maximum takeoff and landing weight is 2315 pounds for R3000140 and 2535 pounds for R300-160. Wingspan is 32 feet 2 inches, overall length is 24 feet 8 inches and overall height is 8 feet 9 inches. Propeller diameter is 6 feet and propeller ground clearance is 12 inches. Empty weight is 1323 pounds for the -140 and 1433 for the -160. Maximum level speed at sea level is 135 knots or 155 mph for -140 and 146 knots or 168 mph for the -160.

ROBINSON HELICOPTER [USA]

Robinson Helicopter Company was founded in June 1973 in Torrance, California by Frank Robinson. They now have 115 factory authorized dealers and 300 service centers in over 55 countries. Helicopter manufacturing, parts and accessories, helicopter and engine overhaul, helicopter repairs and educational courses for pilots and maintenance personnel are their main services. They employ 1200 employees with sales exceeding $200 million annually. The first flight of the R22 two seat prototype was August 1975. In October 1979 they delivered the first R22. By January they had delivered 100 helicopters. They are powered by one 160 HP Lycoming O-320- B2C flat-four engine mounted on the lower rear section of the main fuselage with a cooling fan. It has a light aluminum fuel tank holding 19.2 gallons. The main rotor diameter is 25 feet 2 inches and tail rotor is 3 feet 6 inches. Skid track is 6 feet 4 inches and maximum takeoff and landing weight is 1370 pounds. Cruising speed at 75 percent power at 6000 feet is 96 knots or 110 mph. Range with auxiliary fuel is 319 nautical miles or 368 miles and endurance is 3 hours 20 minutes at 65 percent power and no reserves.

The Robinson R44 is a four place light helicopter. The first production three were delivered In February 1998. It has one 260hp Lycoming O-540 flat-six engine de-rated to 225hp at takeoff and 205hp continuous. It has heating and ventilation. The rotor diameter is 33 feet and the tail rotor is 4 feet 10 inches. Skid track is 7 feet 2 inches. Maximum takeoff and landing weight is 2400 pounds. Cruising speed at 75 percent power is 113 knots or 130 mph. Maximum range with no reserves is 400 miles. The tri-hinged main rotor eliminates lag hinges and hydraulic struts. In February 2005 they opened a new manufacturing facility, which added 220,000 square feet making their total area 480,000 square feet. In April 2005, they delivered the 6,000th helicopter.

ROCKWELL CORPORATION [USA]

Rockwell, located in Seal Beach California was founded in

1928 as North American Aviation and manufactured aircraft. Between 1935 and 1945 North American supplied more than 42,000 military aircraft which included the famous P-51 Mustang fighter and the T-6 Texan trainer. Both first flew in 1940. After World War II, North American developed the first swept-wing fighter, the F-86 Super Sabre, which was the first fighter to fly at supersonic speeds for sustained periods. General Motors sold their controlling interest in 1948 and North American diversified by becoming involved in development of rocket systems and atomic energy. In the 1950's they built the X-15 rocket powered research aircraft for the U.S military and NASA. TheX-59 first flew in 1959 and the X-15 set separate unofficial altitude and speed records during the 1960's. It achieved an altitude of 67 miles, and speed of Mach 6.7. or 6.7 times the speed of sound.

In 1967, North American merged with Rockwell Standard Corporation to form North American Rockwell Corporation, which was later renamed to Rockwell International Corporation in 1973. Rockwell International Corporation's last plane was the B-1 Lancer Bomber first flown in 1984. In the 1960's and 1970's the North American division was the key development center for the Apollo project including the Saturn V rocket's second and final assembly of the entire launcher. They also built the Apollo Command and Service modules. In 1972 they began development of the Space Shuttle for NASA, building five operational orbiters. Their Rocketdyne division (established as part of North American Aviation in 1955) developed the rocket engines used in many space programs, including those for the three stages of the Saturn V rocket and the main engines of the shuttle orbiter. Rockwell International sold its Aerospace and defense units to Boeing in 1996. In 2001, it announced plans to spin off its avionics business and rename them to Rockwell Automation.

ROGERSON HILLER CORPORATION [USA]

In 1942 Stanley Hiller organized Hiller Industries in Berkeley,

California. In 1944 it was renamed United Helicopters Inc. with Hiller and Henry J. Kaiser of Palo Alto, California. In1952, the name was changed to Hiller Helicopters. In 1961, they were purchased by Eltra Corporation, which was a holding company. In 1964 the title was changed to Fairchild Hiller Corporation. In 1973 they were renamed Hiller Aviation. In 1984 Fairchild Hiller assets were sold to Rogerson Aircraft Corporation and named Rogerson-Hiller of Port Angeles, California.

In 1994, Hiller Aircraft Corporation was formed when Jeffrey Hiller, son of Stanley, led an investment group to repurchase assets from Rogerson-Hiller Corporation.

Hiller's model 360, was re-designated as the OH 23 Raven in mid 1962. The OH-23A was an ambulance version with two external mounted side litters It first was powered by one Franklin O-335-4, load capacity of 840 pounds. Five were purchased by the Air Force for trials. Constant improvements were implemented, the H-23C was a three-place, one piece canopy, metal rotor blades and powered with a 250 hp Lycoming VO-540-23B engine. The H-23D was sold to the Army as a basic trainer. It had an improved rotor and transmission system in 1956. In 1962, the H-23F was produced and powered with a Lycoming 305hp VO-540-A1B engine with a range of 225 miles and service ceiling of 15,200 feet. The H-23 G was a 3- place dual control helicopter with a load maximum of 1041 pounds.

Hiller H-32 Hornet had two Hiller 8RJ12B ramjets. It was designated as HO-5 and H-5 by the Military It was the first certified helicopter to use jet engines as YH-32. The HJ-1 had a rotor diameter of 23 feet, as with all these series.

ROMASHKA [RUSSIA]

Romashka Agro-technical Company is located at Veronezh, Russia The Romashka BSKhS is a single seat agricultural aircraft. It is a conventional strut braced low wing monoplane with un-swept, constant-chord wings and tail plane, and swept back fin.

It is a tail wheel type and has non-retractable landing gear with cantilever main wheel legs. High set enclosed cockpit for optimum forward view over the sloping nose. There is a hopper between engine and windscreen and it has spray bars aft of wing trailing edge. The landing gear is standard tricycle type with a tail wheel It is powered by one 355 hp VMBK M-14P nine cylinder, air-cooled radial engine and a three-blade fixed-pitch propeller. Wingspan is 41 feet 2 ½ inches, length is 28 feet 3 ½ inches and overall height is 9 feet 10   inches. The hopper volume is 41.6 cubic feet. Empty weight is 2416 pounds and maximum takeoff weight is 4828 pounds. Maximum level speed is 129 knots or 149 mph, nominal cruising speed is 92 knots or 105 mph and minimum flying speed is 51 knots or 59 mph. Takeoff speed is 8 knots or 67 mph and landing speed is 42 knots or 48 mph. Service ceiling is 13,285 feet, takeoff run is 545 feet, landing run is 475 feet and range with maximum fuel is 275 nautical miles or 317 miles.

SABRELINER [USA]

Sabreliner located in St Louis, Missouri is probably the most known as the original equipment manufacturer of military and private jet aircraft which were manufacture until 1981. The corporation has been privately owned since 1983 when a group of investors purchased the assets from the Sabreliner Division of Rockwell International. This company continues to grow, both by diversification and by acquisition.

In 1977, final assembly of Sabreliner aircraft was moved from Los Angeles to the Perryville, Missouri plant. After transitioning, the Perryville plant produced 8 Sabreliner 60 models, 10 Sabreliner 80 models, and 76 Sabreliner 65 models. All Sabreliner 65s were produced at the Perryville plant. The total production of Sabreliners was 346 airplanes. This plant is located 90 miles south of St. Louis, Missouri and is the home of production and maintenance activities. They keep pace with the industry by performing modifications and major maintenance, as well as other corporate and government aircraft. This facility encompasses 12 buildings consisting of 210,000 square feet under roof. The employees average 21 years of experience. More than 60 percent are licensed Aircraft and Powerplant [A&P] mechanics. Most Sabreliners regularly return to the Perryville facility for maintenance and modification.

Sabreliner-Independence located at Independence, Missouri specializes in engine accessory overhaul and repair. This facility occupies 25,000 square feet dedicated to fuel controls, fuel and oil pumps, valves, igniters, and similar components for commercial and military engines, including General Electric CF700/CJ610, Textron-Lycoming LTS-101 and Pratt and Whitney PT6 and RR250.

SCHWEIZER AIRCRAFT CORP. [USA]

Schweizer was organized in 1939 at Elmira, New York to produce sailplanes. Since that time, they made Grumman Ag-Cat under subcontract, rights to manufacture Hughes 300 light helicopters and supporting Hughes 300 light helicopters and U.S. Army TH-55 trainers. Schweizer subcontracts include Bell Helicopter, Boeing, Sikorsky, and others. They are involved in prototyping and in projects to develop heavy lift vehicles, aerial application for pheromones, centrifuges, and spatial disorientation trainers.

The Schweizer SA 2-37A, Military designation RG-8A was a two-seat special mission surveillance aircraft for the U.S. Army and U.S. Coast Guard. It is powered by one 235 hp Textron-Lycoming IO-540-W3A5D flat-six engine driving a McCauley three-blade constant speed propeller. Standard fuel capacity is 52 gallons increasable to 67 gallons. The wingspan is 61 feet 6 inches and fuselage length is 27 feet 9 inches. Propeller diameter is 7 feet 2 inches and Empty weight is 2,025 pounds. Maximum takeoff weight is 3500 pounds and maximum mission payload is 750 pounds.

The Schweitzer 333 is the pride of the company. It is designed for training, personal use, and utility applications. It is equipped with one Allison 250-C20 turbofan engine. It costs about 30 percent less and operates for about half as much as other single turbines. It has a three-bladed rotor system and allows the 333 to retain a large power reserve to recover the aircraft and to perform emergency procedures. The 333 has the lowest noise signature of any conventional tail rotor turbine helicopter. Its performance is comparable to the Bell Jet Ranger but at a fraction of the cost.

SIKORSKY [USA]

THE MAN: Igor Ivanovich Sikorsky was born in Kiev, Russia on May 25, 1889. His parents, Ivan and Zinaida Sikorsky, were physicians giving the son, Igor, the scientific background to pursue his early dreams of flight. He built a rubber band powered model helicopter when he was 12 years old. Later a larger model with two propellers rose a few feet in the air. Igor entered the polytechnic Institute of Kiev in 1907. He studied aviation, including Count Von Zeppelin and the Wright Brothers, and remained focused on a flying machine that would rise above the ground with a lifting propeller. In a German hotel room on a family trip, he took lift measurements on a four-foot diameter helicopter rotor. With financial backing from his sister he returned to Paris to study aerodynamics and buy components for his helicopter.

In France, early aviation pioneers told the young engineer not to waste his time on the helicopter. Igor returned to Kiev in 1909 with a three cylinder Anzani motorcycle engine and built a helicopter with coaxial twin-bladed rotors. The crude test bed had a seat for the pilot and wires to change the pitch of the blades. He overcame initial vibration problems to demonstrate rotary wing lift. His measurements showed the helicopter could generate about 357 pounds of lift, but 100 pounds less than its empty weight. He abandoned his first helicopter in October 1909 and returned to Paris to study the promise of the airplane.

On return to Russia, Igor first built two propeller driven sleighs. In February he used the engines on a second unsuccessful helicopter and the S-1 small pusher airplane which never flew. The S-2 tractor biplane and larger S-3 lifted off briefly. The S-5 with a 50hp engine flew in ay, 1911 . Mr. Sikorsky was granted FAI pilot license by the imperial Aero Club in Russia. He participated in Russian military maneuvers near Kiev in September 1911 and proved his S-5 was faster than foreign aircraft then used in Russia.

The S-6 with a 100 hp engine was flown in November 1911. In 1912, Igor became chief engineer for the aircraft factory of the Russian Baltic Railroad Car factory in Petrograd. His S-6B won a small interest from the Russian Army and the governing society approved the construction of a large four-engine airplane. It had a wingspan of 89 feet and a gross weight about 9000 pounds. Mr. Sikorsky's S-21 was the Grand. When it first flew on May 13, 1913, Igor Sikorsky became the world's first four-engine pilot. The larger S-22 was dubbed the II' ya Moromets and in December 1913 began flying passengers. A Bomber version flew in 1914 and was used in the war with the Imperial Russian Air Force in 1915.

The Bolshevik Revolution drove him his position and his homeland in 1918. A brief wartime March 30, 1919. After a temporary engineering job ended with the U.S Army Air Service in Dayton, Ohio, he taught mathematics to fellow immigrants on New York's East Side. The slump in aviation development after World War I kept him from his true passion until March 1923 when he received backing for an all-metal, twin-engine passenger plane. After a lifetime of challenges, achievements and honors Igor Sikorsky died in October 1972 at the age of 83.

THE CORPORATION: Sikorsky Aero Engineering Corporation was founded on a farm in 1925 near Roosevelt Field, Long Island, by Igor Sikorsky. In 1929 Igor purchased land in Stratford, Connecticut. Since 1929, it has been a division of United

Technologies Corporation and began helicopter production in the 1940s. Sikorsky's S-38 was sold to ten airlines and Col Charles Lindbergh inaugurated airmail service between the United States and the Panama Canal Zone. Igor's S-40 American Clipper was flown in 1931. The larger and more efficient S-42 Clipper was used by Pan American to open routes across the Pacific and Northern Atlantic. The smaller S-43 succeeded on shorter routs. The Sikorsky line of flying boats ended with the VS-44A Excalibur with seats for 40 passengers.

Sikorsky acquisitions include Schweizer Aircraft in Elmira, New York, Keystone Helicopter Holdings, Inc. of West Chester, Pennsylvania and PZL Mielec, Poland. The service companies include Helicopter Support, Inc. which provides commercial operators the full range of complete factory-authorized services for the S-92, S-76, S-70 and S-61 helicopters, Sikorsky Support Service which is divided into the Aircraft Service Division performs major aircraft modification and crash damage repairs of Sikorsky products, The Maintenance Service Division with responsibility for performing contract maintenance and base support on all fixed and rotary wing aircraft, the Field Service Operations which provides field service representatives for all Sikorsky products; and Derco Aerospace, Inc. of Milwaukee, Wisconsin which is a leader in military aircraft logistics and component distribution, component repairs and aftermarket program management.

Military helicopters are a large portion of the Sikorsky Corporation. The most significant are summarized below. The YR-4B (VS-316A) was one or two-seat combat rescue and general utility helicopter powered by one 200 hp Warner R-550 or a 3 Scarab seven-cylinder air cooled radial engine driving a three-bladed main rotor with a 38 ft diameter. The maximum speed was 75 mph, cruising was 60 mph, service ceiling was 8,000 feet, range of 100 miles and combat radius of 50 miles. It weighed 2020 pounds, normal takeoff weight was 2200 pounds and gross weight was 2535 pounds. First flight was on January 14, 1942.

The next was the H-5 (S-51) Dragonfly, a two to four-seat combat rescue and utility helicopter, powered by a Pratt and Whitney R-985-1/5 Wasp Junior radial fan cooled engine driving a 49-foot diameter three-blade main rotor. Maximum speed was 103 mph, cruise of 75 mph, range of 180 miles and duration of 4 hours. It weighed 3788 pounds empty, normal takeoff weight was 4400 pounds and gross weight of 4985 pounds. First flight was August 18, 1943. The H-5G was an upgraded model and first flight was February 16, 1946.

Sikorsky S-55 was the Army H-19, Navy HO4S and Marine HRS Chickasaw. It was powered with a Wright R-1300-3 air cooled, radial engine driving a four-blade 53 foot diameter main rotor. Maximum speed was 112 mph, cruising speed 91 mph, range of 360 miles, combat radius 100 miles and endurance 5 hours 20 minutes. Empty weight was 5250 pounds, normal takeoff 7200 and maximum takeoff of 7900 pounds. It could carry ten to twelve armed troops or four litters. First flight was November 10, 1949. It was also used by the Navy for anti-submarine roles. This HO4S carried a crew of three, two pilots and one Sonar operator.

Sikorsky S-58 (Army H-34) helicopter was a four-place, two pilots and two Crewmen, for general purposes. It was powered by one Wright R-1820-84 or –84B Cyclone 9-cylinder air cooled radial engine rated at 1525 hp at 2800 rpm, driving a four-blade 56 foot diameter rotor on fully articulated rotor head. Maximum speed was 120 mph, service ceiling 9500 feet, and range 225 miles. Empty weight was 8230 pounds, gross weight was 13,300 pounds and useful load of 5070 pounds. Armament consisted of a door mounted M60 or M1 machine gun. The payload was 3600 pounds of personnel of equipment. The first flight was September 20, 1954 [production]. This helicopter was used also a part of the U.S. Presidential fleet.

The S-76 is a very versatile helicopter both in the military and commercial arena. It is powered by two Pratt and Whitney Canada PT6B-36 engines each with 981 to 1033shp, driving a four-blade main rotor with a diameter of 44 feet. Length overall is 52 feet 6 inches with rotors turning. Crew is two pilots with accommodations for 12-13 passengers. As with all other Sikorsky helicopters it is equipped with a hoist for rescue. Empty weight is 6641 pounds maximum takeoff weight is 11,700 pounds. Maximum speed is 155 knots [166 mph] Range at 139 knots with no reserve is 410 miles and with 30 minute reserve 331 miles. First flight was FAA certification in early 1987. Customers include China, Germany, Japan, South Korea, Netherlands, UK and USA. The first flight of the S-76C was on May 18, 1990. It was upgraded and engines were Arriel 1S1 turbo-shafts. Most other characteristics of the S-76B were retained. First deliveries were April 1991.

The S-80, military designation of MH-53E Super Sea Stallion is a heavy lift transport helicopter. It is designed for two pilots and a crew of four. It has three 3695shp General Electric T-64-GE-416 turbo-shaft engines driving a seven-blade rotor 79 feet in diameter. Maximum speed is 196 mph at sea level. Cruising speed is 173 mph. Ferry range is 1290 miles, operational radius of 575miles with 20,000 pounds external cargo. Empty weight is 33,326 pounds, maximum takeoff 69,750 pounds with an internal load or 73,500 pounds with an external load. First production flight was December 1980. This helicopter is also used by the U.S. Navy in anti-submarine warfare missions.

Sikorsky S-92 Helibus is an outgrowth of S-70 Black hawk/Seahawk designs. The S-92C is designed for offshore oil support, passenger and cargo transport and SAR. The Navy version S-92M has folding tail boom and main rotor blades. The USMC is looking forward for the S-92M to replace the aging CH-46. It is basically Black Hawk rotor head with bifilar vibration damper, inclined crossbeam four-blade tail rotor on right hand side of the pylon, improved main rotor blades with compound swept 30 degrees and drooped 20 degree tips, High set tail plane to left of pylon, SH-60B Seahawk heavy duty transmission, and 44 percent commonality with Black Hawk/Seahawk in rotors and drive trains. The power is supplied initially with two GE CT&-6X and later with two Rolls-Royce Turbo-meca RTM 322 engines, generating between 1750 and 2200shp. Fuel capacity is 580 gallons in sponsons and it has a crash-safe suction fuel system. The main rotor diameter is 53 feet 8 inches and tail rotor is 11 feet in diameter. It has internal floatation, emergency lighting and electrical systems Two pilots are seated on a separate flight deck, with room for 19-22 passengers, with an escape hatch at the rear row of seats. Weight empty is 13,730 pounds. Maximum takeoff weight is 22,220 pounds. Maximum cruising range is 160 knots or 184 mph. Best range speed is 140 knots or 161 mph, Service ceiling is 15,000 feet and range at best speed and 10% reserves is 460 miles. Sikorsky remains on the forefront, and they are researching the Tilt Rotor concept to possibly integrate in future helicopter designs.

UH-60M is the designation of the modernized UH-60 Black Hawk [Navy designation Seahawk]. During the next 25 years, the modernization of 1200 is projected. The U.S. Navy helicopter fleet is expected to be all Sikorsky by 2010, with the MH-60S and MH-60R the only air platforms utilized. Sikorsky has facilities in other Connecticut locations as well as in Alabama, Florida, New York, Pennsylvania, Texas, Wisconsin, and in Poland. Revenues in 2006 were $3.2 billion, employing about 13,000 worldwide. Also, Sikorsky helicopters are in use by all 5 U.S. military forces, as well as military and commercial operators in 40 nations.

SUKHOI [RUSSIA]

This company was named for Pavel Osipovich Sukhoi who was the manager from 1939 until his death in September 1975. It remains one of two primary Russian centers for development and production of fighter and attack aircraft and is expanding activities to include civil aircraft and in cooperation with Western nations. Sukhoi Su-17, Su-20 and Su-22 are named by NATO as Fitter-C, D, E, F, G, H, J. and K.

Su-17M is a single seat attack aircraft with a AL-21F-3 engine, eight stores pylons, additional wing fence on each glove panel, curved dorsal fin and operational with CIS air forces and naval aviation. Su-17R is a reconnaissance version. Su-17M-2/M-2D is Fitter D and is basically the same as Su-17M, but forward fuselage is lengthened by 15 inches and drooped 3 degrees to improve pilots view while keeping the intake face vertical, added under-nose Doppler navigation radar pod, and a laser rangefinder is in the intake center body. Su-17UM-3 is Fitter G. It is a two-seat trainer version of Su-17M with combat capability, drooped forward fuselage, deepened dorsal spine fairing for additional fuel , taller vertical tail surfaces, removable ventral fin, starboard wing root gun only, and laser range finder in the intake center body.

Su-17M-3 is Fitter H, which is an improved single seat airplane with a gun in each wing root and launcher for R-60 air-to-air missile between each pair of under wing pylons. The Fitter H/Ks are equipped for tactical reconnaissance carry, typically, centerline sensor pod, active ECM pod is under the port wing glove, and two under wing fuel tanks.

Su-17M-4 is a single seat version with cooling air intake at front of dorsal fin. Otherwise it is the same as Su 17M externally. The Su-20 (Fitter C) is an export version of Su-17M. Su-20R is the export version of Su-20 Some export aircraft are equipped with Tumansky R-29 BS engines with 25,350 pound st afterburning in a more bulged rear fuselage, rearranged small external air intakes at the rear fuselage and shorter plain metal shroud terminating fuselage.

Su-22U (Fitter E) is a two-seat trainer developed from Su-17M-2, with Taumansky engine, no Doppler pod, deepened dorsal spine fairing for additional fuel and the port wing root gun is deleted. Su-22 (Fitter F) is the export version of Su-17M. It has modified under nose electronics pod, R-29 engine, gun in each wings root, weapons include R-3 (AA-2Atoll) air-to-air missiles. Aircraft sold to Peru had limited coverage radar warning system and almost no navigation aids. Some U.S. supplied avionics. Su-22UM-3K (Fitter G) is the export version of Su-17UM-3 with AL-21F-3 or R-29B engine.

SU-22M-4 (Fitter K). These were provided to Russian CIS air forces and Naval Aviation, as well as Afghanistan, Algeria, Angola, Czech Republic, Egypt, former East Germany, Hungary, Iraq, North Korea, Libya, Peru, Poland, Slovakia, Syria, Viet-Nam and Yemen. The landing gear is retractable tricycle type, with a single wheel on each unit. Nose wheel retracts forward and main landing gear retracts inward into the center section. The nose wheel is steerable + or − 27 degrees, and there is a container for single cruciform brake-chute between the base of the rudder and tailpipe. The power plant is one Saturn/Lyulka AL-21F-3 turbojet with 17,200 pounds st and 24,800 pounds st with afterburner. Fuel capacity is 1212 gallons with the same capacity in dorsal spine fairing. There are provisions for drop tanks on outboard wing pylons and under the fuselage. When under fuselage tanks are carried, only the two inboard wing pylons can be used for ordnance to total 2204 pounds. There are two solid-propellant rockets units attached optionally to rear fuselage to shorten the takeoff run. It is heavily armed and updated as technology advances. Wingspan fully spread is 44 feet 10 1/2 inches and fully swept is 32 feet 10 3/4 inches. Overall length including probes is 62 feet 5 inches and fuselage is 52 feet 1 inch long. Weight empty, equipped is 23,737 pounds. Maximum external stores weigh 8820 pounds and maximum takeoff weight is 41,887 pounds. Maximum landing weight is 28,660 pounds. Maximum speed at altitude is Mach 1.74, 1000 knots or 1150 mph, and sea level speed is Mach 1.10, 729 knots or 838 mph. Service ceiling is 46,585 feet and takeoff run is 2955 feet. Landing run is 3120 feet. Maximum range at high altitude is 1242 nautical miles or 1430 miles, and at low altitude, 556 nautical miles or 870 miles.

Su-24 is a two engine, two seat variable geometry battlefield bomber reconnaissance and FW aircraft. The NATO name is Fencer. Fencer A has rectangular rear fuselage box enclosing jet nozzles. Modifications result in Fencer B (Su-24) first operational airplane. It has a deep-dished bottom skin to rear fuselage box between jet nozzles, and larger brake-chute housing at the base of the rudder. Fencer C had significant avionics changes, multiple nose fitting instead of a simple probe, triangular fairing for RWR on the side of each engine air intake forward of fixed wing root and each side of the tip, chord of fin leading edge is extended forward, except at tip giving a kinked profile.

Fencer D is a major attack version. It is believed to have terrain-following radar instead of the earlier terrain-avoidance system. Also it has inflight refueling capability, with centrally mounted retractable probe forward of the windscreen, nose is about 2 feet 6 inches longer to accommodate a new avionics bay, large over-wing fences with integral extended wing root glove pylons when carrying KH-29 (NATO name Kedge) missiles. Under nose antennae was deleted, laser rangefinder/designator housing aft of nose wheel bay, and a single long nose probe. The empty equipped weight was 49,162 pounds. This is also the export version of Su-24MK.

Fencer E is a reconnaissance version, Su-24MR. It is equipped with side-looking radar in the nose, Aist –M- TV reconnaissance system and panoramic and oblique cameras in the ventral fairing. A laser pod can be carried on the centerline, with a ELINT pod or radiation detector pods on starboard under wing swiveling pylon and two air-to-air missiles under port wing. Data can be transmitted to the ground by data link. It has no over wing fences. Provisions are included for two under wing auxiliary fuel tanks with a capacity of 792 gallons each. Other capabilities include inflight refueling and air-to-surface missiles.

Fencer F, Su-24MP is an electronic warfare/jamming/SINGINT version There is a small fairing under the nose, no underside electro-optics, hokey stick antennae as on Su-24MR and a centerline EW pod. More than 950 units have been delivered to date. Users include the Russian forces, Iraq, Iran, Libya, and Syria. The landing gear is hydraulically retractable tricycle type with twin wheels on each unit. Main units retract forward and inward into air intake fairings. It has a steerable nose wheel retracting rearward, and brakes on the main wheels with anti-skid units, mudguard on the nose wheels and two cruciform brake chutes, each 269 square feet. It is powered by two Saturn/ Lyulka AL-21F-3A turbojets, each producing 24,690 pounds st with afterburning and fixed engine air intakes. Four internal tanks have a capacity of 3090 gallons and can be supplemented by two 528 gallon external tanks under the fuselage and two 792 gallon tanks under wing gloves. It has gravity and pressure fueling and probe-and-drogue flight refueling capability including as a buddy tanker. Aircraft supplied to Iran have 54 chaff/flare dispensers in the wing fence in addition to standard 24 on the sides of the rear fuselage. The armament is constantly being updated as technology advances but we must understand that this airplane is very heavily armed and it is equipped with the most up-to-date countermeasures. Wingspan is 57 feet 10 ½ inches with 16 degree sweep and 34 feet with 69 degree sweep. Overall length including the probe is 80 feet 8   inches. Empty weight is 41,885 pounds, maximum internal fuel is 21,525 pounds, and external stores total 17,635 pounds. Normal takeoff weight is 79,365 pounds, maximum takeoff weight is 87,520 pounds and maximum landing weight is 57,320 pounds. Maximum speed, clean at height is Mach 1.35 and at sea level Mach 1.08, or 712 knots or 820 mph. Stalling speed with flaps and wheels down is 151 knots or 174 mph. Maximum rate of climb at sea level is 29,525 feet per minute and service ceiling is 57,400 feet. Takeoff run is 4265 feet and takeoff to 50 feet altitude is 4920 feet. Landing run is 3120 feet ad landing from 50 feet is 5250 feet. Combat radius is lo-lo-lo is over 174 miles or 200 miles. Lo-lo-hi with 5500 pounds of weapons is 590 miles or 515 nautical miles. Hi-lo-hi with 6615 pounds of weapons and twin external tanks, 565 nautical miles or 650 miles. G limit is +6.5.

SUKHOI Su-25 and Su-28, NATO name Frogfoot, is a single-seat close support aircraft and a two-seat trainer. Su-25 is Frogfoot A and the export version is Su-25K. Su-25UB, Frogfoot B is without weapons. It is two-seat, and is used for operational and weapons trainer. The rear seat is raised, resulting in a humpback appearance. It has a separate hinged portion of the continuous framed canopy over each cockpit, a taller tail fin and increasing overall height. Only a few were built and overall length is 50 feet 4 inches. Su-25UTG [Frogfoot B] is as the Su-UT but with carrier landing equipment such as a arrester [tail hook] under the tail. It is used primarily for landing training on a dummy flight deck marked on the runway at Saki airfield. Ten were built for these trials.Su-25BM had under-wing pylons for rocket propelled targets for release for missile training by fighter pilots. This was the same as SU-25 except for these dummies. Besides being used by Russian tactical forces, it was exported to Afghanistan, Bulgaria, Czechoslovakia, Hungary and Iraq.

Su-25BM has shoulder-mounted wings, anhedral from the roots and extended chord leading edge on outer 50 percent of each wing. Wingtip pods are split at the rear airbrakes that project above and below the pod when extended. Emphasis on survivability features account for 7.5 percent of takeoff weight. This includes an armored cockpit and the fuel tanks are filled with reticulated foam for protection against explosion. The landing gear is a tricycle type, retractable, hydraulically operated. The main single wheels retract to be horizontally in the bottom of the engine air duct. The nose wheel is a steerable single wheel that retracts forward. Two cruciform brake chutes are housed in the tail cone. The two engines are Soyuz/Tumansky R-195 in long nacelles at the wing roots, each producing 9921 pounds st. There are thick armor firewalls between engines. The turbojets have pipe-like filament at the end of the tail cone from which air is expelled to lower temperature and lower the infrared signature. Fuel tanks are located in the fuselage between the cockpit and front wing spar and between the rear spar and fin leading edge and in the wing center section. there are provisions for four external fuel tanks on under-wing ylons. The single pilot seat is a zero-zero ejection, under sideways hinged to starboard canopy with a small rear-view mirror, and a bullet proof windscreen. A folding ladder is for access to the cockpit, is built into the port side of the fuselage. The wingspan is 47 feet 1 ½ inches and overall length is 50 feet 11 ½ inches. Weight empty is 20,950 pounds and maximum takeoff weight is between 32,187 and 38,800 pounds. Maximum landing weight is 29,320 pounds and maximum level sped is Mach 0.8, 526 knots or 606 mph. Maximum attack speed with air brakes open is 372 knots or 428 mph and typical landing speed is 108 knots or 124 mph. Service ceiling clean is 22,965 feet and with maximum weapons is 16,400 feet. Typical takeoff run is 1970 pounds and with weapons maximum from unpaved surface is 3935 feet. Normal landing run is 1970 feet and with brake chutes, 1312 feet. Range with 9700 pounds weapon load and two external tanks at sea level is 405 nautical miles or 466 miles. At altitude, it is 675 nautical miles or 776 miles. G limits with 3306 pounds of weapons is +6.5 and with 9700 pounds of weapons +5.2.

SU-26M is a single seat aerobatic airplane. It is marketed worldwide, including the UK and U.S.A. It is a typical aerobatic competition airplane of symmetrical section, mid-wing. The controls are mechanical. Landing gear is a non-retractable tail wheel type with hydraulic disc brakes. It is powered by one 394 hp nine-cylinder radial engine with a three blade variable pitch propeller. One option is for a two-blade variable pitch propeller. The fuel tank in the fuselage forward of the front spar with a capacity of 16.6 gallons. Tanks in each wing leading edge holds total of 53 gallons for ferry flights. The canopy is designed as rearward hinged with jettisonable capability. The seat is designed for the pilot to use a backpack parachute. The electrical system is 24/28 volt. The wingspan is 25 feet 7 inches and overall length is 22 feet 5 ½ inches. Propeller diameter is 7 feet 10 ½ inches. Empty weight is 1554 pounds and maximum takeoff weight is 2205 pounds. Vne (never exceed speed) is 243 knots or 280 mph, maximum level speed at sea level is 167 knots or 192 mph and normal cruising speed is 140 knots or 161 mph. Takeoff speed is

65 knots or 75 mph and landing speed is 62 knots or 72 mph. Service ceiling is 13,125 feet, takeoff run is 525 feet and landing run is 820 feet. Ferry range at 3280 feet altitude is 432 nautical miles or 497 miles. Operating G limits are +12/-10, and ultimate is +23.

Su-27, NATO name Flanker, is a single seat all-weather air superiority fighter and single/two-seat ground attack airplane, or a two-seat combat trainer. The current version is Flanker B single-seat, land based production with square wingtips, carrying air-to-air missile launchers, ECM jammer pods, extended tail cone and forward retracting nose wheel. Su-UB is a tandem two-seat trainer with full combat capability. The Su-27K is for shipboard use. Su-27KUIB is a side-by-side two-seat version and Su-27KU is a deck landing trainer, but no wing folding or tail hook. Customers include to Russian elements, China and Syria. The controls are four channel fly-by-wire artificial feel with no mechanical backup. It has a automatic angle of attack limiter limiting it to 30 to 35 degrees, but it can be overridden manually. There is a large door-type air brake in the top of the center fuselage. The landing gear is hydraulically retractable tricycle type with a single wheel on each unit. The main gear retracts forward into wing roots. The steerable nose wheel with mudguard retracts forward. It is powered by two turbofan engines with afterburning, each producing 27557 pounds st. A fine grill screen on each inlet duct protect the engine from foreign object damage (FOD) during takeoff and landing. Air conditioning is automatically regulated. The ejection seat is zero-zero type under a large forward opening transparent blister canopy with a low sill. It is equipped with a full array of the latest weapons and countermeasures. wingspan is 48 feet 2 inches and overall length excluding the nose probe is 1 feet 11 ½ inches. Maximum takeoff for Flanker B is 48,500 to 67,240 pounds and for Flanker C 49,600 pounds. Maximum level speed is Mach 2.35, 1350 knots or 1550 mph and at sea level Mach 1.1, 725 knots or 835 mph. Service ceiling is 59,055 feet. Takeoff run is 1640 feet for the Flanker B is

1640 feet and for the Flanker C is 1805 feet. Landing run for the Flanker B is 1970 feet and for the Flanker C is 2135 feet. Combat radius is 930 miles, range with maximum fuel for the Flanker B is over 2160 nautical miles or 2485 miles, and for the Flanker C is 1620 nautical miles or 1865 miles. G limit for both models is +9.

Su-29 is a tandem two-seat training/single-seat aerobatic airplane. It is generally the Su-26M. It has a three-blade propeller and capacity of each wing fuel tank is 28.15 gallons. Total fuel capacity is 72.9 gallons. It has gravity fueling and is single pilot for aerobatic competition and dual for training. Propeller diameter is 8 feet 2 ½ inches. Empty weight is 1698 pounds, maximum fuel is 456 pounds and maximum takeoff weight is 2655 pounds. Service ceiling is 13,125 feet, never-exceed speed is 237 knots or 273 mph, maximum level speed is 175 knots or 202 mph and takeoff run is 394 feet. Landing run is 1247 feet and range with maximum fuel is 647 nautical miles or 745 miles.

Su-30 is a two-seat long-range superiority fighter. It is designed for missions of 10 hours or more and carry bombs and rockets but not guided air-to-surface weapons. Most characteristics are the same as for Su-27. Range with one refueling is 2805 nautical miles or 3230 miles. Su-30MK is a two-seat multi-role fighter. It is completely armed with an array of rockets, missiles, and bombs totaling up to 17,635 pounds. Normal takeoff weight is 52,910 pounds and maximum takeoff weight is 72,750 pounds. Maximum level flight speed is Mach 2.0, takeoff run is 1805 feet and landing run is 2200 feet. Combat range with internal fuel is 1620 nautical miles or 1865 miles and with one refueling 2805 nautical miles or 3230 miles.

Su-31T is a single seat aerobatic competition and training airplane. The basic version has a non-retractable landing gear, but SU-31U has a retractable landing gear. It is basically a version of SU-31T and is constructed with more than 70 percent composites by weight. It is powered by one 395hp nine-cylinder radial engine and a three-blade propeller with a diameter of 8 feet 2 ½ inches. Wingspan is 25 feet 7 inches and overall length is 22 feet 7 inches.

The propeller ground clearance is 1 foot, 4 inches. Empty weight is 1482 pounds, maximum internal fuel is 117 pounds and with external fuel is 461 pounds. Maximum takeoff weight is 2134 pounds, maximum never exceed speed is 243 knots or 280 mph and maximum level speed is 183 knots or 211 mph. Stalling speed is 61 knots or 71 mph, takeoff speed is 60 knots or 69 mph and landing speed is 57 knots or 66 mph., and Service ceiling is 13,125 feet. Takeoff run is 360 feet and landing run is 985 feet. Range with internal fuel is 156 nautical miles or 180 miles, maximum ferry range is 400 to 645 nautical miles or 460 to 745 miles and g limits are +12/-11. Su-32 is a tandem two-seat primary trainer and military general-purpose airplane. It is based on the Su-29 and SU-31 acrobatic airplanes with objectives of short takeoff and landing, high maneuverability, plus carriage of guided and unguided weapons for patrol and coastal protection. Controls are mechanical with option for an autopilot. The landing gear is pneumatically retractable tricycle type with a single wheel on each unit. The main units retract inward and nose wheel retracts rearward. Hydraulic brakes are on the main landing gear. It is powered by one 355hp nine-cylinder engine with a three-blade propeller. The cockpit is pressurized and air-conditioned. The propeller and windscreen are de-iced by hot air. There are provisions for integral gun, bombs, air-to-air and anti-tank missiles in the combat versions. Wingspan is 27 feet 10 inches and overall length is 23 feet 10 7/8 inches, propeller diameter is 8 feet 2 ½ inches and propeller ground clearance is 9 inches. Empty weight is 1874 pounds, maximum internal fuel is 573 pounds and external 441 pounds for a total of 1014 pounds. Maximum takeoff and landing weight is 3307 pounds. Never exceed speed is 286 knots or 329 mph, maximum level speed is 200 knots or 230 mph and maximum cruising speed is 178 knots or 205 mph. Economical cruise is 151 knots or 174 mph. Stalling speed with flaps down is 49 knots or 56 mph and with flaps up, 54 knots or 62 mph. Service ceiling is 13,125 feet. Takeoff run is 492 feet and to 50 feet altitude, 1313 feet. Landing run is 656 feet, and for altitude of 50 feet, 1970 feet. Range with maximum payload is 647 nautical miles or 745 miles and with maximum internal and external fuel, 1080 nautical miles or 1242 miles.

Su-33 NATO named Flanker D. is a single-seat ship-based defense fighter. It is similar to Su-27, but some upgrades and modifications were accomplished. It has folding wings, arrester [tail] hook and other features to adapt it for shipboard operations. The landing gear is strengthened, twin nose wheels and the long tail cone is shortened to prevent tail scrapes during takeoff and landing. The flying controls are basically the same as Su-27 with added 52 degree swept back wings, fly-by-wire automatic control system and central control column and broader chord slotted fowler type flaps. The structure is the same as Su-27 except with hydraulically folding outer wings through 135 degrees and upward folding horizontal tail surfaces. It is a riveted and welded structure of aluminum and titanium alloys and steel. The landing gear is generally the same as Su-27 but larger tires, nose wheel steerable through plus or minus 60 degrees and the arrester [tail] hook is under the tail cone. The power plant is the same as Su-27, except it has a retractable flight refueling probe beneath the windscreen on the port side and has provisions for centerline buddy refueling pack. Wingspan is 48 feet 2 inches and overall length including the probe is 69 feet 6 inches. Never exceed speed, at 36,000 feet is Mach 2.165, 1240 knots or 1430 mph. Minimum flying speed is 130 knots or 150 mph. Takeoff on a carrier with 14 degrees ramp is 395 feet, range with maximum fuel is 1620 nautical miles or 1865 miles and g limit is +8.Su-34 is a long-range theater bomber, planned to replace Mig-27, Su-17 and some SU-24. In general it is the same as Su-271B. Changes include, diaelectric nose to house nav/attack and terrain following/ avoidance radar, tail fins that contain fuel tanks, titanium cockpit armor, new main landing gear with units smaller, tandem wheels and a larger diameter tail cone. It houses, at its tip, a rearward facing radar to detect aircraft approaching from the rear. It may carry a number of rearward facing infrared air-to-air missiles for use against aircraft flying at altitudes from 165 to 42,650 feet over ranges of 3300 feet to 7.5 miles. It is powered by two turbofan engines each producing 30,865 pounds st with afterburning. It accommodates two pilots side-by-side on zero/zero ejection seats. It is armed with an array of missiles, both air-to-air- and air-to-ground, including laser guided. Maximum takeoff weight is 97,800 pounds. Maximum level speed at 36,000 feet is Mach 1.8 and at sea level, Mach 1.15 or 755 knots or 870 mph. Range is 2160 nautical miles or 2485 miles.

Su-35 is an all-weather counter-air fighter and ground attack airplane. It is an advanced version of the Su-27 with three color CRT in the cockpit and engines produce 30,865 pounds st with afterburning. A single pilot sits on a zero/zero ejection seat. Weapon array is updated and the structure is a higher proportion of carbon fiber and aluminum lithium alloy in the fuselage construction and composites are used for components such as leading edge flaps and the nose wheel door. Avionic and the basic Su-27 are updated in all respects. This results in a wingspan of 49 feet 2 ½ inches and overall length is 72 feet 2 inches. Maximum level speed at altitude is Mach 2.35 or 1350 Knots or 1555 mph. At sea level, limits are Mach 1.18, 782 knots or 900 mph. Service ceiling is 59,055 feet altitude. Runway length required is 3940 feet and range with maximum fuel is 2160 nautical miles or 2485 miles and with in-flight refueling more than 3510 nautical miles or 4040 miles.

Su-38 is a single/dual agricultural airplane. It is a conventional low wing monoplane with non-retractable landing gear, similar to Su-29 two-seat acrobatic airplane. The landing gear has a radial engine with a three-blade propeller. Each wing has a fuel tank with a capacity of 53 gallons. Occupants normally include one pilot, but there are provisions for one passenger. Spray bars are mounted under the wings, and chemicals are in either ventral pod or fuselage hopper replacing the passenger seat. Wingspan is 32 feet 9 inches and overall length is 23 feet 10 inches. Propeller diameter is 8 feet 2 ½ inches and ground clearance of the propeller is 1 foot 7 ½ inches. Empty weight is 2139 pounds, maximum payload is 1102 pounds and maximum fuel is 330 pounds. Maximum takeoff and landing weight is 3637 pounds, never exceed speed is 175 knots or 202 mph. Maximum level speed at sea level is 162 knots or 186 mph.. Maximum cruising speed at sea level is 151 knots or 174 mph, economical cruising speed is 65 knots or 75 mph and service ceiling is 9840 feet. Takeoff run is 395 feet, and to 50 feet altitude 985 feet. Landing run is 920 feet and landing from 50 feet is 1640 feet. Range with maximum payload is 377 nautical miles or 435 miles and with maximum fuel it is 540 nautical miles or 620 miles.

Su-39 is a single seat anti-tank aircraft. It is basically the Su-25 with improvements incorporated. Using the lessons learned from t the war in Afghanistan, mainly survival in intense anti-aircraft defense environment. A new navigation system makes it possible for travel to and from combat areas under automatic control. Equipment in the widened nose includes: Television activated about six miles from the target, subsequent target tracking, weapon selection and automatic release, wing tip countermeasures pods incorporated and gun is transferred to under belly position on starboard side of further offset nose wheel. The flying controls are basically the same as Su-25. Trim is only on the rudder and it has artificial feel in lateral and longitudinal channels. It also has an automatic control system. The landing gear has main wheels with metal and ceramic disc brakes with anti-skid units. Two brake chutes, each 140 square feet. The power plant is as the Su-25 and the cockpit is pressurized. It has updated avionics with a laser range finder and target designator that is improved. It has a chaff/flare dispenser on top of the fuselage tail cone and in a large cylindrical housing at the base of the rudder that also contains an infrared jammer, optimized against Stinger and Redeye missile frequencies. It is heavily armed, including guns, missiles that will penetrate 900 mm reactive armor, laser guided missiles, anti-radiation air to-surface missiles, bombs and rockets and air-to-air missiles. Wingspan is 47 feet 7   inches and overall length is 50 feet 4 ½ inches. The maximum combat load is 9612 pounds, maximum fuel is 15,233 pounds and maximum takeoff weight is 42,990 pounds. Maximum landing weight is 29,100 pounds, never-exceed speed is Mach 0.82, maximum level speed at sea level is Mach 0.77, 512 Knots or 590 mph. Economical cruising speed is 350 knots or 404 mph. Service ceiling is 29,525 feet, takeoff run is 1970 feet and 2300 feet on unpaved runway and landing run is 2300 feet. Combat radius with 4410 pounds of weapon load at sea level is 215 nautical miles or 248 miles. At altitude the range is 378 nautical miles or 435 miles. Ferry range is 1350 nautical miles or 1550 miles and g limit is +6.5.

S-21 is a supersonic business airplane It is a three-engine low/mid wing airplane and the wings have compound sweep with highly swept inner panels, an oval section fuselage and all-moving swept foreplanes. Shockwave designed level is 3 to 4 times lower than the Concorde, sounding like remote thunder at sea level. Controls are digital fly-by-wire automatic flight control system. It has three turbofans, each producing 16,535 pounds st and accommodations include a crew of two and 6 to 10 passengers. The wingspan is 65 feet 4 inches, overall length is 124 feet 2 ½ inches. The empty operating weight is 54,167 pounds, normal payload is 1015 pounds, maximum payload is 2006 pounds and maximum fuel is 58,465 pounds. Maximum takeoff weight is 114,200 pounds. Maximum supersonic cruising speed is Mach 2.0, 1150 knots or 1320 mph, and subsonic Mach .95, 547 knots or 630 mph. Cruising altitude is 50,850 to 63,975 feet. Takeoff and landing run is 6500 feet and range is 4000 nautical miles or 4600 miles.

S-51 is a medium long-range supersonic transport. It has rear mounted sweptback low wings with very long curved root extensions and a small-diameter long fuselage. It is designed to fly from Moscow to any city in the Asian and European continent, or non-stop to any world capital while being permitted to fly at supersonic or subsonic speeds above highly populated areas. It is powered by four turbojets producing 20,945 pounds st each. It is intended to carry a crew of two and from 50 to 60 passengers. The wingspan is 92 feet 10   inches and overall length is 165 feet 8 inches. Operating empty weight is 86,060 pounds, maximum payload is 13,670 pounds, maximum fuel is 103,670 pounds and maximum takeoff weight is 199,955 pounds. Maximum supersonic speed is Mach 2.0, 1145 knots or 1320 mph and subsonic 547 knots or 630 mph. Service ceiling is 55,775 to 62,335 feet and runway requirement is 8200 feet. Maximum range is at Mach 1.4 and reduced fuel is 2930 nautical miles or 3375 miles. At Mach 2.0 or .95 it is 4965 nautical miles or 5715 miles. Noise levels are reduced to comply with all standards.

S-54 is a two seat jet trainer and light combat airplane. It is a scaled down Su-27 development. The goals of this project are to improve speed, operating altitude and maneuverability commensurate with combat aircraft. It has an un-conventional all swept wing, twin outwards canted fins mounted at the wing trailing edges, engine air intakes are under the wing roots and retractable tricycle landing gear. The controls are fly-by-wire as the Su-27 by way on flap irons, all moving tail plane and rudders, leading edge flaps and the airbrake is pre-programmable to make airplane easier or harder to fly dependant on student's ability. Optional "panic button" is available to return the aircraft to straight and level flight from any altitude and pushbutton spin recovery, and optional playback to record every move of the student in flight. The landing gear is retractable with the main wheels retracting inward and the nose wheel retracting forward. It has no brake/chute. It is powered by one turbojet 13,670 pounds st with after burner. Two internal fuel tanks have a total capacity of 3660 pounds with single point pressure fueling. It accommodates two pilots in tandem under a blister canopy. The seats are zero/zero ejection seats with the rear seat raised to improve visibility. It is armed with two wing tip Infrared homing air-to-air missiles and two hard points under each wing for mounting air-to-air and/or air-to-surface weapons.

The wingspan is 26 feet 9 ½ inches and overall length is 40 feet

4 inches. Empty equipped weight is 10,560 pounds and maximum takeoff weight is 20,745 pounds. Maximum landing weight is 15,718 pounds. Maximum level speed at height is Mach 1.55, or 890 knots or 1025 mph. At sea level, it is Mach .98, or 645 knots or 745 mph. Takeoff speed is 98 knots or 112 mph, landing speed is 92 knots or 106 mph. Service ceiling is 59,050 feet, takeoff run is 1250 feet and landing run is 1640 feet. Range at sea level with maximum fuel is 510 miles or 440 nautical miles and at height 1080 nautical miles or 1240 miles. The g limits are +9 and –3.

S-80 is a twin turbo prop multi purpose STOL transport. It was the largest and most advanced program of the former Soviet industry. It was manufactured under several versions. The current version is S-80P, which was a basic cargo/passenger version, to carry up to 24 passengers or 5510 pounds of cargo. Other versions are S-80DT, a military light assault airplane, S-80GR adapted to carry geological exploration equipment, S-80M for ambulance to carry 10 stretcher patients and attendant, S-80PT a patrol or transport version and S-80R, a fishery surveillance version. It is powered by two turboprop 1480 shp engines, each driving a six-blade reversible thrust propeller. Two fuel tanks are provided with a total capacity of 660 gallons. There is an APU that may be operated at remote and/or unprepared sites. It has a retractable tricycle type landing gear, the main twin wheels retract into the tail booms. The nose gear is steerable and retracts rearward. It is designed for a crew of two seated side-by-side and 19 to 23 passengers or 10 stretcher patients or the equivalent freight or equipment. New avionics are provided as well as advanced flight deck equipment.

Wingspan is 76 feet ½ inch and length is 54 feet 8 inches. Propeller diameter is 8 feet 8½ inches, ground clearance of the propellers is 3 feet 7 inches and distance between centers of the propellers is 18 feet 4 ½ inches. Empty weight is 13,735 pounds, operating weight empty is 14,330 pounds, maximum payload is 5511 pounds, and maximum fuel is 5225 pounds. Maximum takeoff weight is 24,250 pounds and maximum landing weight is

22,925 pounds. Never exceed speed is 296 knots or 341 mph, maximum level speed at 19,685 feet is 270 knots or 310 mph, economical cruise at 19,685 feet is 194 knots or 223 mph and service ceiling is 26,250 feet. Takeoff run is 1180 feet, takeoff to 50 feet altitude is 2725 feet, landing from 50 feet is 2495 feet and landing run with reverse thrust is 590 feet. Range with maximum payload is 675 nautical miles or 776 miles and with maximum fuel is 2425 nautical miles or 2795 miles.

S-84 is a four or five-seat multi-purpose airplane. It is a cantilever mid wing monoplane with winglets, flush canopy, upswept at the rear, sweptback fin and T tail plane, swept back ventral fin of the tail cone. The airframe is all composites with two spar wings. The landing gear is retractable tricycle type with a single wheel on each unit and the nose wheel is steerable plus or minus 40 degrees. It is powered by one 350 hp Teledyne Continental TSIOL-550B flat-six piston engine in center of the fuselage behind wings, driving a five-blade propeller. Fuel capacity is 153 gallons. The avionics are designed for Bendix/ King Silver Crown for VFR and IFR, day or night into paved or unpaved surfaces including those without ILS aids. Options include a bar, audio vision equipment, TV and satcom. Wingspan is 41 feet, 4 inches, overall length is 31 feet 10 inches. Propeller is 8 feet 6 ½ inches in diameter and propeller ground clearance is 3 feet 5 inches. Empty weight is 2436 pounds, maximum payload is 1102 pounds, maximum fuel is 948 pounds and maximum takeoff and landing weight is 4188 pounds. Never exceed speed is 235 knots or 270 mph, maximum level and maximum cruising speed at 19,685 feet is 205 knots or 236 mph, and maximum cruising speed at 13,125 feet is 118 knots or 136 mph. Stalling speed with flaps down is 51 knots or 59 mph and service ceiling is 19,685 feet. Takeoff run is 1115 feet, to 50 feet altitude is 1575 feet. Landing run is 590 feet and maximum range with maximum payload is 1370 nautical miles or 1578 miles. It is 1570 nautical miles or 1808 miles with maximum fuel.

S-86 is a 7-seat twin-turboprop multi-purpose transport. It is a unconventional high wing with sweep forward of high aspect ratio at the rear of the fuselage with center section with much increased chord extended rearward to carry widely separated twin tail units. Single main wheels are retractable and retract into large sponsons. It is powered by two 542 shp turboprop engines driving pusher counter rotating propellers through a combining gearbox The engine air intake is flush on each side of the rear fuselage. The cabin is pressurized, and provisions made are for alternative missions such as cargo, medical and patrol.

Wingspan is 52 feet   inch, and overall length is 37 feet 11 ½ inches. Empty operating weight is 6395 pounds, maximum fuel is 2976 pounds, maximum payload is 1190 pounds and maximum takeoff weight is 9920 pounds. Maximum level speed is 323 knots or 373 mph and service ceiling is 34,450 feet. Takeoff and landing from 50 feet altitude is 1640 feet. Range with maximum fuel is 1885 nautical miles or 2175 miles.

S-986 is a twin-engine multi-purpose light transport airplane. It is a high wing pod fuselage, twin boom design with winglets, inward canted swept back fins and rudders. The landing gear is hydraulically retractable tricycle with a single wheel on each unit. Main wheels retract rearward into the booms. It is powered by two 355hp, nine cylinder radial engines, driving three-blade propellers. Fuel capacity is 145 gallons. It accommodates a crew of two side-by-side. The cabin will seat up to nine passengers or it may be equipped for agricultural forest patrol or emergency rescue. The wingspan is 52 feet 6 inches and overall length is 38 feet and   inch. The propeller diameter is 8 feet 2 1/2 inches, distance between propeller centers is 16 feet 1 inch. Empty weight is 5070 pounds, maximum fuel is 1212 pounds. and maximum takeoff weight is 8818 pounds. Never exceed speed is 242 knots or 280 mph, maximum level speed is 188 knots or 217 mph, and economical cruising speed is 113 knots or 130 mph. Service ceiling is 13,125 feet, takeoff run is 2050 feet and to 50 feet altitude is 2625 feet. Landing run is 1150 feet and to 50 feet altitude 1725 feet. Range with maximum fuel is 810 nautical miles or 932 miles.

THURSTON AEROMARINE CORP. [USA]

The Thurston Aeromarine Corporation is located at Foreside, Maine. The International Aeromarine Corporation in Sanford Florida ceased production during 1991. Design data and all rights to TA16 Seafire rights reverted to Thurston Aeromarine Corporation. The Thurston TSC-TSC-1A3 Teal III is a two-seat amphibian. It is similar to the TA16, except it is powered by one Textron Lycoming O-360-A1F6D 180 hp engine instead of the originally planned 150hp engine The hull section provides to accept tricycle landing gear with nose wheel steering. Maximum takeoff weight is 2300 pounds. The propeller is a Hartzell HC-C2YK-1BF/F7666A-2 tractor propeller. Two leading edge fuel tanks contain 46 gallons. It has enclosed cabin seating two side-by-side and baggage area behind the seats. Empty weight is 1500 pounds, and maximum takeoff weight is 2300 pounds. Maximum level speed is 101 knots or 116 mph, cruising speed at 75 percent power is 97 knots or 112 mph Service ceiling is 16,000 feet, endurance is 5 and ½ hours and g limit is 5.7. The Thurston TA16 Seafire made its first flight on 10 December 1981, with deliveries in 1993. It is powered by one Textron Lycoming O-540-A4D5 flat-six 250 horsepower engine. The propeller is a Hartzell two-blade constant-speed metal tractor type. Fuel capacity is 90 gallons in the leading edge of each wing. A refueling point is on the upper surface of each wing. The engine and propeller are pylon mounted on the upper surface of the hull directly over the wings. Wingspan is 37 feet and the hull is 24 feet 4 inches. The cabin provides four seats. Empty weight is 1950 pounds and maximum takeoff and landing weight is 3200 pounds. Range with maximum fuel is 1000 miles. Service ceiling is 18,000 feet and takeoff from land or water requires 1200 feet and landing requires 1200 feet.

Thurston TA19 Seamaster is an eight seat Transport Amphibian. Production deliveries began in 1997. It was developed as a turboprop replacement for the Grumman Goose and Mallard Amphibians. Accommodations are for two pilots and 8 passengers. Wingspan is 53 feet and length overall is 39 feet. Empty weight is 4860 pounds and maximum takeoff weight is 8600 pounds It has a retractable tricycle landing gear with steerable nose wheel, underwing floats and skis are optional. Power is supplied by two 450 shp Allison 250-B17F turbo-shaft engines, each driving a Hartzell three-blade, constant speed propeller. Four interconnected fuel tanks in each wing have a total capacity of 360 gallons. Service ceiling is 20,000 feet, cruising speed at 75 percent power is 174 knots or 200 mph. Takeoff distance for land is 140 feet and sea is 1850. Landing is 1200feet for land and 1400 feet for water. Range with maximum standard fuel, 50 percent power and with 30 minute reserve is 1340 miles. Range with maximum payload, power and 30 minute reserve is 1100 miles. Endurance with maximum standard fuel is 9 hours.

TUPOLEV [RUSSIA]

This company was founded in 1918 and named after and led by Andrei Nikolayevich Tupolev who died in 1972 at the age of 84. It concentrated on large military and civil aircraft until the early 1990's. Current effort is 80 percent civil efforts. No new military programs were projected as of 1994. TU-16, named by NATO Badger is a twin jet medium bomber, maritime reconnaissance, attack and electronic warfare airplane

It has several versions, the most current Tu-16A and Badger A. Other versions are: Tu-16KS was the first version with air-to-surface missile, which is currently retired. Tu-Z was an experimental flight refueling tanker using unique wingtip-to-wingtip techniques. Tu-T torpedo bomber. Tu-16S Korvet, adapted for search and rescue missions with a large radio controlled rescue boat under the fuselage. Tu-16K-10 (Badger D) is an anti-ship airplane using a turbojet powered missile, a wide nose radome, and no provision for free fall bomb. Tu-16 (Badger F) is for photographic/electronic reconnaissance but has cameras in weapons bay and two additional radomes under the fuselage a larger one aft. Tu-16R was similar to Badger E but Elint pod under each wing, and later versions have small radomes under the center fuselage. Tu-16K-11/16 badger C was a conversion of Tu-16KS with two underwing rocket powered missiles which can be carried over a range of greater than 1735 nautical miles or 2000 miles. This version was delivered to Naval forces and the free falling bombing was retained. Tu-16K-26, Badger G modified to carry a Mach 3 missile with nuclear or conventional warhead under each wing, a large radome replacing the chin radome. Tu-16K-10-26, Badger C modification, modified to carry two Kingfish missiles under the wing in addition to Tu-10S capability. Tu-16N is equipped as a flight refueling tanker and added tanks in the bomb bay. Tu-16PP, Badger J, is a standoff or escort ECM airplane with the primary function of chaff dispensing to protect missile carrying strike forces. It is specialized for active jamming of all frequencies. Tu-16R, Badger K, is an electronic reconnaissance version with the nose same as Tu-16A. Tu-16B Badger L, is a naval electronic warfare like Badger G, but with equipment of the kind that is on the Tu-95 [Bear G] including an ECM nose thimble, extended tail cone containing specialized equipment. instead of the tail gun. Tu-16 LL is a test bed for other engines and equipment. Tu-16 Tsiklon is a meteorological laboratory and Tu-16-G was used by Aeroflot for urgent mail transportation. The design is an all-swept high mid-wing airplane with heavy engine nacelles from root fairings, 42-degree sweepback at the quarter chord on the tail fin and tail plane and the tail plane incidence is one degree and 30 minutes. The landing gear is Hydraulically retractable tricycle type, twin wheel nose wheel retracts rearward, main gear has four-wheel bogies that retract into housings projecting beyond the wing trailing edge. The power plant consists of two 20,920 pounds st turbojets, semi-recessed in the sides of the fuselage. Fuel contained in 27 wing and fuselage tanks have a total capacity of 11,570 gallons. There are provisions for under wing auxiliary fuel tanks for flight refueling. Tankers trail the hose from starboard wing tip and receiving is in the port wingtip extension. It can be very heavily armed, including a bomb load of 19,800 pounds in the weapons bay. Naval versions may carry air-to-surface winged standoff missiles. Wingspan is 108 feet 3 inches and overall length is 114 feet 2 inches. Empty equipped weight is 82,000 pounds, maximum fuel is 75,750 pounds, and normal takeoff weight is 165,350 pounds. Maximum landing weight is 110,230 pounds. Maximum level speed at 19,700 feet altitude is 567 knots or 652 mph and service ceiling is 49,200 feet and range with 6600 pounds of bombs is 3110 nautical miles or 3580 miles.

Tu-95 and Tu-142, NATO name Bear, is a four turboprop long range bomber and maritime reconnaissance airplane. It is powered by four 12,000shp engines. The Tu-142 is an anti-submarine version. The Tu-95 and Tu-142 went through several modifications to achieve their goals. They are unique high performance aircraft with all-swept high aspect mid wing design. The fuselage is the same diameter as the U.S. B-29 and Soviet Tu-4. The main landing gear retracts into the wing trailing edge nacelles, the contra-props have high tip speeds. The landing gear has retractable tricycle type, main units consisting of four wheel bogies and twin wheel steerable nose wheels. They all retract rearward, main units into nacelles constructed on to wing trailing edge. The power plant is four 14,795 shp engines, driving eight-blade contra-rotating reversible pitch propellers. Fuel tanks in wings have a normal capacity of 25,100 gallons and they have a flight-refueling probe for extended ranges. They carry a crew of seven with no ejection seats. The accommodations are pressurized, and anti-icing of wings and tail plane leading edges. They have an APU in the dorsal fin with the exhaust above the tail plane leading edge. Life raft stowage is in front of the dorsal fin, and they have eight, three-tube flare/chaff dispensers in two rows aft of landing gear doors on the lower side of each wing pod. Wingspan is 167 feet 8 inches overall length is 162 feet 5 inches and propeller diameter is 18 feet 4 ½ inches. Empty weight is 198,415 pounds for the Bear H and maximum fuel is 191,800 pounds for the Bear F. Maximum takeoff weight for Bear F is

407,850 pounds and for the Bear H is 414,470 pounds. Maximum landing weight for Bear B is 297,620 pounds. Bear H maximum level speed at sea level is 350 knots or 404 mph and at 25,000 feet altitude Mach 0.83, 499 knots or 575 mph. At 38,000 feet it is Mach 0.78, 447 knots or 515 mph. Normal cruising speed is 384 knots or 442 mph. Takeoff speed is 162 knots or 187 mph and approximate landing speed is 146 knots or 168 mph. Service ceiling is normally 39,370 feet and with maximum weapons. 29,850 feet. Combat radius with 25,000 pounds payload is 3455 nautical miles or 3975 miles without refueling, and with one in-flight refueling, 4480 nautical miles or 5155 miles. Tui-22, NATO name Blinder, is a twin jet supersonic bomber and maritime reconnaissance airplane. It has some versions as follows: Tu-22K, Blinder B, is a basic bomber with weapon doors redesigned to carry AS-4 Kitchen air-to-surface missiles which are semi-recessed. It has a larger radar and partially retractable flight-refueling probe on the nose. Tu-22R, is a daylight reconnaissance version with six windows in the weapons bay doors for three pairs of long focal-length cameras. It has a chaff dispensing chute aft of the weapons bay and a flight refueling probe. On some aircraft the tail gun is substituted by ECM equipment. A modification to this version named Blinder C Mod, extends the capability for night reconnaissance through addition of a centerline conformal pack, possibly containing Infrared and ECM and possible ESM just forward of the wing on each side. Tu-22U, Blinder D is a training version with a raised cockpit for the instructor aft of the standard flight deck with stepped up canopy.

Tu-22P, Blinder E is an electronic warfare version with avionics and cooling systems in the weapons bay. It is an upgrade of all weapons and, intelligence and countermeasures. It is an all swept mid-wing design with engine nacelles mounted above the rear fuselage on each side of the fin with intake lips. Main landing gear units retract into the pods on the wing trailing edges. The leading edges are swept back 52 degrees. The fuselage is basically a circular section with area ruling at the wing roots. It is loaded with ECM and ECCM equipment including chaff/flare dispenser and bombing assessment cameras. It carries a crew of three in tandem. The pilot has an upward ejection seat and other crewmembers have downward ejection seats. The controls are hydraulically powered. The landing gear is retractable tricycle type, wide track with four main wheel bogie units retract rearward into pods built onto wing trailing edges. The nose wheel retracts rearward and a retractable rear skid is to protect the rear fuselage in tail down landing or takeoff and a twin brake chute is standard.

Wingspan is 77 feet 1 inches and overall length is 139 feet 9 inches. Maximum fuel load is 93,695 pounds, maximum weapons load is 26,455 pounds and normal takeoff weight is 187,390 pounds. Maximum takeoff weight is 202,820 pounds and with JATO 297,230 pounds. Normal landing weight is 132,275 pounds. Maximum level speed at 40,000 feet is Mach 1.52, 870 knots or 1000 mph. Landing speed is 168 knots or 193 mph and supersonic service ceiling is 43,635 feet. Takeoff run is 7385 feet, landing run is 7120 feet and landing run with brake-chutes is 5415 feet. Radius of action is 700 to 1188 nautical miles or 807 to 1365 miles. Range with maximum fuel is 2645 nautical miles or 3045 miles and ferry range is 3050 nautical miles or 3510 miles.

Tu-22M, NATO name Backfire, is a twin engine variable geometry maritime reconnaissance/attack airplane. The current version is Tu-22M-2. Backfire B is the first production series. Tu-22M-3, Backfire C, is the advanced long range bomber and maritime version. A possible version may be for Electronic Warfare. The surfaces are all swept back and it is capable of delivering nuclear weapons. The landing gear is hydraulically retractable tricycle type, with rear-retracting twin nose wheels. Each main wheel bogie comprises three pairs of wheels in tandem with varying distances between each pair pivot inward from the vestigal fairing under the center section on each side into bottom of fuselage. The brake/chute is located inside a large door under the rear fuselage. It is powered by two 55,115 pounds st turbofan engines with afterburners. Fuel tanks in the center fuselage between engine ducts carry through the structure. An APU is mounted in the dorsal fin and provisions for JATO are included. The pilot and co-pilot are seated side by side under upward opening jettisonable gull wing doors hinged on the centerline. Ejection seats are provided for all four crewmembers. It is heavily armed with the most up-to-date weapons. Wingspan fully spread is 112 feet 5 inches and fully swept 76 feet 5 ½ inches. Overall length is 139 feet 3 inches. Maximum weapon load is 52,910 pounds, fuel load is 110,230 pounds, maximum takeoff weight is 273,370 pounds and maximum takeoff weight with JATO is 278,660 pounds. Normal landing weight is 171,955 pounds and maximum landing weight is 194,000 pounds. Maximum level speed at high altitude is Mach 1.88, 1080 knots or 1242 mph and at low altitude it is Mach 0.86, 567 Knots or 652 mph. Normal cruising speed at height is 485 knots or 560 mph. Takeoff speed is 200 knots or 230 mph, normal landing speed is 154 knots or 177 mph and service ceiling is 43,635 feet. Normal takeoff run is 6560 to 6890 feet, and normal landing run is 3940 to 4265 feet. Non-refueled combat radius ranges from 810-1000 nautical miles or 930-1150 miles to 1300 nautical miles or 1495 miles and maximum g level is+2.5.

Tu-160, NATO named Blackjack, is a four-engine variable geometry long range strategic bomber. For comparison it is about 20 percent longer than USAF B-1B with greater un-fueled combat radius and greater maximum speed. The controls are fly-by-wire and the structure of the fuselage is slim and shallow and shaped for hostile radar signal non-detection. The landing gear consists of twin nose wheels that retract rearward. The main wheels consist of two bogies, each with three pairs of wheels. Retraction is similar to Tu-154 airliner. Each leg pivots rearward and the bogie rotates through 90 degrees around the axis of the center pair of wheels parallel with the retracted legs, into the thickest section of wings between fuselage and inboard engine on each side. It is powered by four 50,580 pounds st turbofan engines with afterburning. It is equipped with an in-flight refueling probe that retracts into the top of the nose. It accommodates a crew of four on individual ejection seats. The controls are fighter-type sticks rather than yokes or wheels. Armament, with a maximum of 36,000 pounds, consist of free fall bombs, short range attack missiles, or ALCMs. It contains no guns. Wingspan fully spread is 182 feet 9 inches, with 35 degree sweep 166 feet 4 inches and fully swept 116 feet 9 inches. Overall length is 177 feet 6 inches. Empty weight is 242,500 pounds, maximum fuel is 352,735 pounds and maximum weapon load is 36,000 pounds. Maximum takeoff weight is 606,260 pounds and maximum landing weight is 341,710 pounds. Maximum level speed at 40,000 feet is Mach 2.05, 1200 knots or 1380 mph. Cruising speed at 45,000 feet is Mach 0.9, 518 knots or 596 mph and service ceiling is 49,200 feet. Takeoff run is 7220 feet and landing run with maximum weight is 5250 feet. Radius of action at Mach 1.5 is 1080 nautical miles or 1240 miles. Maximum un-refueled range is 6640 nautical miles or 7640 miles and g limit is +2.

Tu-24SKh is a single/two seat agricultural monoplane. It has a non-retractable tail wheel and the main gear is carried on side Vs and half axles. The cabin is designed for one or two people side-by-side and a hopper in the cabin aft of seats. It is powered by one nine-cylinder air cooled radial engine and a two-blade propeller. Wingspan is 42 feet 8 inches and overall length is 30 feet 4 inches. Maximum chemical weight is 1985 pounds and maximum takeoff weight is 5630 pounds. Maximum level speed is 127 knots or 146 mph, normal cruising speed is 108 knots or 124 mph, spraying speed is 65 to 75 knots or 75 to 87 mph. Takeoff run is 590 feet and landing run is 328 feet. Ferry range is 1080 nautical miles or 1240 miles.

Tu-34 is a STOL multi purpose light airplane. It has retractable tricycle type landing gear with a single wheel on each unit. Main wheels retract inward and upward into the fuselage. It is designed for operation on un-prepared airfields. It is powered by two 220hp piston engines. Accommodations are for a pilot and four passengers, or 992 pounds of cargo, and is convertible for ambulance and patrol missions. Wingspan is 43 feet 3 inches and overall length is 30 feet 1 inches. Maximum takeoff weight is 4188 pounds. Maximum level speed is 215 knots or 248 mph, normal cruising speed is 151-183 knots or 174-211 mph and service ceiling is 26,900 feet. Balanced runway length is 1315 feet and range with maximum fuel is 1133 nautical miles or 1305 miles.

Tu-130 is a twin turboprop short-range passenger and freight transport. It is a conventional high wing monoplane with the wing in three sections. They are un-swept except the tail fin which is swept back. The landing gear is retractable tricycle type, a twin wheel nose unit and each main unit has two wheels in tandem retracting into long fairings on sides of the fuselage. It is powered buy two 2465shp turboprop engines. Provisions are included for the engines to operate on a butane/propane mixture and liquefied natural gas. Accommodations are provided for up to 53 passengers with galley and toilet. Alternative uses are for freight or vehicles, including four large pallets. Wingspan is 87 feet 1 inch, overall length is 74 feet 7 inches and rear-loading ramp is 105.5 square feet. Maximum payload is 11,023 pounds and maximum takeoff weight is 46,300 pounds. Normal cruising speed is 270-280 knots or 310-323 mph, normal cruising height is 22,965 feet, balanced runway length is 5905 feet, range with maximum payload is 1080 nautical miles or 1242 miles and with 6615 pounds of cargo, 2160 nautical miles or 2485 miles.

Tu-160SC is a conversion of Tu-160 bomber as space launch vehicle. It can carry Burlak Two-stage rockets under the fuselage on the centerline mount. Design payload into orbit is planned at a height of 124-620 miles; 1765-1323 pounds into polar orbit; and 2425-1850 pounds into equatorial orbit. It carries a crew of four. The Burlak rocket is 70 feet 2 ½ inches long, has a diameter of 4 feet 1 inches and the Burlak launch weight is 66,135 pounds. Maximum takeoff weight is 606,260 pounds. Speed for launch at 29,500 to 36,000 feet is Mach 0.8 and at 30,000 to 42,650 feet is Mach 1.7. Normal cruising speed is Mach 0.77, 459 knots or 528 mph and required runway length is 11,500 feet.

Tu-154M and Tu-154S, NATO name Careless, are three turbofan, medium range transports. The basic Tu-154M is a basic airliner for up to 180 passengers. Tu-154M2 is a modernized version with PS-90A turbofans that consume 62 percent as much fuel per passenger. Life is projected to be 20,000 hours and 15,000 cycles Tu-154S is a special freight version with unobstructed main cabin cargo volume of 2,542 cubic feet. Nominal range is 1565 nautical miles or 1800 miles with 44,100 pounds of cargo. The landing gear is hydraulically actuated retractable tricycle type. Main wheel bogies each have three pairs of wheels in tandem retractable rearward into fairings on wing trailing edge. The anti-shimmy twin nose wheel retracts rearward and is steerable +63 degrees or –63 degrees, and the main wheels have disc brakes and are anti-skid. They are powered by three 23,380 pounds st turbofans in a pod on each side of the rear fuselage and inside of the extreme rear of fuselage. The two lateral engines have clamshell thrust reversers. The wings have integral fuel tanks, four tanks are in the center section and two are in the outer wings. All fuel is fed to a collector tank in the center section and then to the engines, and they have single point refueling. They accommodate a crew of three, two pilots and flight engineer with provisions for a navigator and five cabin attendants. There are two passenger cabins separated by a bulkhead. It provides for 180 tourist class passengers with hot meal service, 164 tourist class with hot meal service, or 154 tourist/economy plus first class seating eight to 24 persons. Seating is mainly six-abreast with a center aisle. The entire airplane except a hold under the rear of the cabin are pressurized and air-conditioned. Each has a fire extinguisher in the engines and smoke detectors in the baggage holds. Wingspan is 123 feet 2 ½ inches and overall length is 157 feet 1 inches. Basic empty operating weight is 121,915 pounds, maximum payload is 39,680 pounds and maximum fuel is 387,633 pounds. Maximum takeoff weight is 220,460 pounds and maximum landing weight is 176,366 pounds. Maximum cruising speed is

513 knots or 590 mph and maximum cruising altitude is 39,000 feet. Field length required is 8200 feet, maximum range with maximum payload is 2105 nautical miles or 2425 miles, with 26,455 pounds of payload 2805 nautical miles or 3230 miles and with maximum fuel and 12,015 pounds of cargo, 3563 nautical miles or4100 miles.

Tu-204 is a twin turbofan medium range airliner. It is designed to carry up to 214 passengers or maximum payload of 46,295 pounds. The versions are as follows: Tu-204-100 is an extended range version with additional fuel. Maximum takeoff weight is 227,070 pounds; Tu-204-120 is the same, but with Rolls Royce engines. Tu-204C is a cargo version, Tu-204-200 has a further increase in payload and additional fuel, Tu-204-220 has Rolls Royce turbofan engines with 43,100 pounds st. Tu-204-222 is the same as the basic but with Collins avionics, Tu-204 –230 is the basic but with Pratt & Whitney turbofans producing 41,700 pounds st. 204-30 is a trunk route version with a shorter fuselage. There are short and mid range models for 166 passengers and a long range model for 99 to 160 passengers. It is equipped with 35,580 pounds st turbofan engines. It is a conventional low/mid wing design with all surfaces swept back. The life is planned for 45,000 flights or 60,000 flight hours Controls are triplex fly by wire with analog backup. Typical Y yokes were selected. About 18 percent of the airframe is made of composites. The landing gear is hydraulically retractable tricycle type with steerable twin wheel nose unit and retracts forward. The main gear is made up of four wheel bogie units that retract inward into the wing/fuselage fairings. It has carbon disc brakes. The power is supplied by two turbofan engines in composite cowlings as described in the versions. Wingspan of the –204, 204-200 and –204-220 is 137 feet 9 ½ inches. For –204-300 short range, midrange and long range is 134 feet 11/2 inch.. Overall length is 150 feet for -204, 204-200 and 204-220 is 150 feet 11 inches and for the others is 131 feet 103/4 inches. Empty operational weight is between 128,530 and 130,070 pounds. Maximum payload is between 39,682 and

55,555 pounds, maximum fuel is between 52,910 and 72,09 and maximum takeoff weight is between 186,950 and 244,155 pounds. Range with maximum payload ranges from 1800 miles to 4475 miles, and with the design payload, 2110 miles to 5750 miles. Tu-244 is a supersonic airliner. It has four turbofan engines, each producing 72,750 pounds st and nominal seating for 300 passengers Wingspan is 178 feet 8 ½inches, and overall length is 291 feet. Empty weight is 2379,180 pounds and maximum takeoff weight is 771,605 pounds. Nominal cruising speed at 59,000 to 65,600 feet altitude is Mach 2.05 and maximum range is 4965 nautical miles or 5715 miles.

Tu-330 is a twin turbofan transport. The landing gear is retractable tricycle type with a twin nose wheel and each main unit has three pairs of wheels in tandem. Wingspan is 142 feet 9½ inches and overall length is 137 feet 9 ½ inches. Design payload is 77,160 pounds and maximum takeoff weight is 227,070 pounds. Nominal cruising speed at 36,000 feet is 431-458 knots or 497-528 mph. Range with 44,090 payload is 3020 nautical miles or 3480 miles and with 66,135 pounds of payload is 1620 nautical miles or 1865 miles.

Tu-334 is a two-turbofan or twin prop-fan medium range airliner. The Tu-334-100 is the turbofan version and carries 72 two-class passengers or 102 tourist class passengers. The prop-fan version is lengthened for 104 to 137 passengers, and has a slightly reduced wingspan and height. The pusher prop-fans engines are mounted on the rear of the fuselage and each produces between 17,635 and 19,840 pounds st. The controls are fly-by-wire with mechanical backup for the tail plane. Composites make up 20 percent of the structure by weight. The landing gear is tricycle type with twin wheels on each nit. The main wheels retract inward into wing/fuselage fairings. The prop-fan version accommodates a crew of two or three n the flight deck and a fourth seat for an instructor or observer. The passenger seat arrangements include 86 seats with 8 seats four-abreast in first class, 12 business class six abreast and 66 tourist class seats six-abreast; 92 seats with 8 first class or 102 all tourist seats. Wingspan is 97 feet 8 inches for the turbofan and 95 feet 5 inches for the prop-fan. Overall length is 102 feet 6 inches for the turbofan and 121 feet inch for the prop-fan. Empty weight for the turbofan is 90,500 pounds, maximum payload is 24,250 pounds for the turbofan and 29,760 pounds for the prop-fan. Maximum fuel for the turbofan is 21,030 pounds. Maximum takeoff weight for the turbofan is 101,630 pounds and for the prop-fan is 103,615 pounds Normal cruising speed at 34,775 to 36,40 feet altitude is 431-442 knots or 497-510 mph for the turbofan and 431 knots or 497 mph for the prop-fan. Runway length is 7220 to 7545 feet for bath, and range for the turboprop with 20,395 pounds of payload or 102 passengers is 1080 nautical miles or 1242 miles.

Tu-414 is a twin turbofan long-range business airplane and regional transport. It is an all Swept, low wing monoplane with a T tail-plane. The controls are conventional and the landing gear is retractable tricycle type with twin wheels on each unit. It is powered by two 14,900 pounds st Rolls Royce turbofan engines.

VOUGHT AIRCRAFT COMPANY [USA]

The original Lewis and Vought Corporation was founded in 1917 and became Chance Vought in 1960. They merged with Ling-Temco Electronics in 1961 forming Ling-Temco-Vought and later it was renamed the LTV Corporation. The main output included about one-third of Northrop-Grumman airframe by weight, Boeing 747 fuselage and tail surfaces, 767 tail plane, tail surfaces of the 757, engine nacelles and tail sections of McDonnell Douglas C-17A, and engine nacelles for Canadair Challenger and Regional Jet. Their most recent is the Vought (FMA) Pampa 2000 International is a tandem seat jet primary trainer, and reengineering of the Eurocopter AS 565 Panther named the Vought [Eurocopter] Panther 800.

YAKOVLEV [RUSSIA]

This company is located in Moscow and was founded by Alexander Sergeyevich Yakovlev, one of the most versatile designers in the Soviet Union, who died in 1989 at the age of 83. One of the first was the Yak-3 which was a reproduction of the World War II wood skinned single seat piston engine fighter which flew in 1943. This was the metal skinned airplane. It was powered by a reconditioned 1240hp Allison 12 cylinder V liquid cooled piston engine which replaced the 1300hp engine with a three-blade propeller. Wingspan was 30 feet 2 inches and overall length was 27 feet 10 inches. Maximum takeoff weight is 5945 pounds, maximum level speed at altitude is 350 knots or 402 mph and at sea level 307 knots or 354 mph. Time to turn 360 degrees is 19 seconds.

Yak-18T is a four-seat multi-purpose light airplane. It is a conventional cantilever low wing monoplane. It is all-metal constructed except for fabric covered outer wing panels, ailerons and tail surfaces. The landing gear is tricycle single wheel on each unit with pneumatic retraction. Main wheels retract inward into the center section and equipped with oleo shock absorbers. The caster nose wheel is not steerable. Main gear has pneumatic brakes. The engine is one 355 hp radial piston engine with a two-blade variable pitch metal propeller Wing root fuel tanks have a capacity of 55 gallons. The cabin seats four persons in pairs. The ambulance configuration accommodates a pilot, stretcher patient and medical attendant.

Wingspan is 36 feet 7 inches and overall length is 27 feet 6 ½ inches. Maximum payload with instructor and one student is 675 pounds and with four passengers 960 pounds. Maximum takeoff weight is 3307 pounds for the training version and 3637 for four persons. Maximum level speed is 159 knots or 183 mph, maximum cruise is 135 knots or 155 mph, and service ceiling is 18,000 feet. Takeoff run is 1085 feet for the training version and 1315 feet for the passenger version. Landing run is

1315 feet for the training version and 1640 feet for the passenger version. Range for the training version is 350 nautical miles or 403 miles and for the passenger version 485 nautical miles or 560 miles.

Yak-42 is a three-turbofan short or medium range passenger transport. The other versions are minor except Yak-42T is a freighter with a maximum of 26,455 pounds including the standard eight containers in the under floor holds. The main design concerns were basic design, simple construction, reliability for operation, economy, and ability to operate in remote areas with wide variations in climate. It is conventional all swept low wing construction with two turbofan engines on the sides of the rear fuselage. The flight controls are hydraulically actuated and the structure is all metal box wing, riveted, bonded and welded semi-monocoque fuselage. The landing gear is hydraulically retractable heavy duty. It has twin nose wheels which retract forward. The main wheel is a four wheel bogie that retracts inward into the fuselage. It is powered by three three-shaft turbofans producing 14,330 pounds st. It has no thrust reversers, and integral fuel tanks between spars in the wings hold 6120 gallons. It accommodates a crew of two side-by-side with provisions for a flight engineer and two or three attendants. The cabin seats 120 passengers six abreast. An alternative is to seat 104 passengers, which are made up of 96 tourist and eight first class. This is a local passenger version with baggage and coat stowage compartments fore and aft of the cabin. An APU is included which provides power and air conditioning supply on the ground and if necessary in flight. All passenger and crew accommodations are pressurized, air conditioned and provided with non-flammable materials. The APU is standard for engine starting, and for power on the ground and if necessary in flight. Wingspan is 114 feet 5 inches and overall length is 119 feet 4 inches. Empty weight is 76,058 pounds for 104 seats and 76,092 for 120 seats. Maximum payload is 28,660 pounds, maximum ramp weight is 126,320 pounds, maximum takeoff weight is 125,660 pounds and maximum landing weight is 112,433 pounds. Maximum cruising speed is 405 knots or 503 mph, economical cruise is 405 knots or 466 mph. Takeoff speed is 119 knots or 137 mph, approach speed is 114 knots or 131 mph and maximum cruising height is 31,500 feet. Field takeoff requires 7220 feet and landing from 50 feet altitude is 3610 feet. Range at economical cruise and with 6615 pounds of fuel reserves and maximum payload is 745 nautical miles or 857 miles, with 120 passengers1080 nautical miles or 1243 miles, with 104 passengers 1240 nautical miles or 1430 miles and with maximum fuel and 42 passengers, 2215 nautical miles or 2545 miles.Yak-242 is a twin turbofan short/medium passenger transport. It superseded the three engineYak-42M and it is a fly-by-wire concept. The landing gear is hydraulically retractable tricycle type. There are four four-wheel bogies for the main wheels and twin nose wheels. It is powered by two turbofans each producing 26,525 pounds st. It accommodates a crew of two seated side by side with provisions for a flight engineer if required. Passenger load accommodates 130 to 180. One option is single class cabin for 180 passengers or 12 four-abreast in first class at the front, 24 business class and 96 tourist seats. Wingspan is 118 feet 11 inches and overall length is 124 feet 8 inches. Operating weight equipped is 84,655 pounds, normal fuel is 30,865 pounds, maximum fuel is 48,500 pounds, maximum payload is 39,680 pounds and maximum takeoff weight is 142,415 pounds. Normal cruising at 36,400 to 38,050 feet altitude is 431-458 knots or 497-528 mph. Runway requirement is 7220 feet and range is1163 nautical miles or 994 miles with maximum payload, 1457 nautical miles or 1677 miles with normal payload and 2700 nautical miles or 3107 miles with maximum fuel.

Yak-44E is a twin turboprop airborne early warning and control airplane. It is similar to the U.S. Grumman E-2 Hawkeye. Due to security, information is scant. The wings fold upward and forward, and it is equipped with a 23 foot 11 ½ inch radome mounted on top of the rear fuselage, and the height overall with the radome up is 23 feet and with the radome down the height is 16 feet 10 1/2 inches. Wingspan is 82 feet 8 inches and overall length is 67 feet 6 inches. The landing gear has two main wheels. Propellers are contra rotating on long shafts. It is equipped with two tail fins and the radome lowers to the same height of fins and folded wings for stowage. Yak-46-1 is a two turbofan short/medium transport. The versions projected are: (1) Single class with seating for 168; (2) Mixed class for 12 first class and 114 tourist passengers; and (3) conversion for passenger/freighter or casualty evacuation. The controls are fly-by-wire and the landing gear is tricycle type, retractable, with four-wheel bogie main units and twin nose wheels. It is a low-wing monoplane with under wing engines. Wingspan is 118 feet 11 inches and overall length is 127 feet 3 ½ inches. Empty weight is 76,808 pounds, maximum payload is 3,580 pounds and maximum takeoff weight is 132,715 pounds. It accommodates a crew of two side-by-side and provisions for a flight engineer if required, and 150 seats six-abreast with a center aisle and galley, toilets and seats for flight attendants front and rear. The power plant includes two high bypass turbofans, each rated at 24,250 pounds st. Normal cruising speed is 448-458 knots or 515-528 mph and normal cruising altitude is 36,400 feet. Runway length required is 6890 feet and range with maximum payload is 1187 nautical miles or 1367 miles, or with normal payload 1863 nautical miles or 2143 miles.

Yak-46-2 is a twin prop-fan short or medium transport. The accommodations and avionics are the same as Yak-46-1. The power plant is two prop-fan engines, each rated at 24,690 pounds driving contra-props, one with eight blades and the other with six blades. The engines are pushers on each side of the rear fuselage on pylons. Wingspan is 116 feet 5 inches and overall length is 134 feet 6 inches. Propeller diameter is 12 feet

5 inches. Empty weight is 82,230 pounds, maximum payload is 38,580 pounds and maximum takeoff weight is 135,140 pounds. Nominal cruising speed is 448 knots or 515 mph, nominal cruising altitude is 36,400 feet and runway length required is 6890 feet. Range with maximum payload is 971 nautical miles or 1118 miles and with normal payload is 1888 nautical miles or 2175 miles.

Yak-54 is a tandem two-seat sporting and aerobatic training airplane. It has a tail wheel type landing gear and is made of all metal, two-spar wings, semi monocque fuselage, a conventional tail unit and a titanium spring main landing gear. It is powered by one 355hp nine cylinder air-cooled engine and a three-blade variable pitch propeller. It has two seats in tandem under a continuous transparent canopy. Wingspan is 26 feet 91/4 inches and overall length is 22 feet 8 inches. Maximum takeoff weight with one pilot is 1874 pounds and with two occupants 2182 pounds. Never exceed speed is 243 knots or 280 mph, stalling speed is 60 knots or 69 mph, ferry range is 377 miles or 435 miles, and g limits are +9 and −7.

Yak-55M is a single seat acrobatic competition airplane. It is mid wing configuration, non-retractable tail wheel landing gear and titanium spring main landing gear legs and a rear sliding canopy. It is powered by one 355 hp nine cylinder air-cooled radial engine, and a two-blade controllable pitch propeller and wing fuel tanks with a capacity of 31.5 gallons. Wingspan is 26 feet 6 inches and overall length is 22 feet 11 ½ inches. Maximum takeoff weight is 1852 pounds and maximum level speed is 243 knots or 280 mph. Stalling speed is 57-60 knots or 66-69 mph and g limits are +9 and −6.

Yak-58 is a six-seat multi purpose airplane. It has a pusher 355 hp nine-cylinder radial Engine, enclosed in an annular duct and a three-blade variable pitch propeller, reducing noise in the cockpit. The tail is a twin fin type on each side of the engine. The wings are un-swept with dihedral tips and the fuselage is mounted above the wings. The landing gear is retractable tricycle with single wheels and low-pressure tires. Main wheels retract inward and the nose wheel retracts forward. When seats are removed, the cabin can be used for freight. Wingspan is 41 feet 8 inches and overall length is 28 feet ½ inch. Empty weight is 2800 pounds, maximum payload is 992 pounds, and maximum takeoff weight is 4630 pounds. Maximum level speed is 162 knots or 186 mph, maximum cruising speed is 153 knots or 177 mph, landing speed is 68 knots or 78 mph and service ceiling is 13,125 feet. Takeoff run is 2,000 feet, landing run is 1970 feet and range with maximum fuel and 45 minutes reserve fuel is above 540 nautical miles or 620 miles.

Yak-77 is a twin turbine long-range executive and regional transport. It is an all-swept low wing airplane, and has rear mounted turbo-fan pods. The landing gear is retractable tricycle type with twin wheels on each unit. The two engines are Allison GMA 3000 turbofans It accommodates a crew of two side-by-side on the flight deck and the cabin accommodates 8 business passengers or 32 for a regional transport. Lengthened versions can accommodate 50 to 70 passengers. The wingspan is 70 feet 8 ½ inches and overall length is 67 feet1 inches. Maximum payload is 7715 pounds, executive version payload is 1985 pounds and maximum takeoff weight is 55,555 pounds. Maximum cruising speed is Mach 0.8 and economical cruising speed at 40,000 feet altitude is Mach 0.75. Runway length required is 7220 feet and maximum range with the eight-seat version is 5400 nautical miles or 6215 miles. The 32 seat version range is 3240 nautical miles or 3725 miles.

Yak-112 is a four-seat multi-purpose airplane. It is intended to accommodate passengers, light cargo and mail, training, glider towing, ambulance, pipeline and cable patrol, fisheries surveillance and agricultural missions. It is high wing, un-swept; the tail fin is swept and composites are used extensively. The landing gear is non-retractable, single wheel low pressure tires on each unit, cantilever spring main gear legs and floats and skis are optional. Power is supplied by a 210 hp six cylinder Teledyne Continental IO-360-ES, driving a Hartzell two-blade, two-position propeller. It accommodates four persons in pairs in an enclosed cabin. The wingspan is 33 feet 7 ½ inches and overall length is 22 feet 10 inches. Empty weight is 1709 pounds, maximum payload is 595 pounds and maximum takeoff weight is 2778 pounds. Maximum cruising speed is 135 knots or 155 mph, economical cruising speed is 102 knots or 118 mph, landing speed is 69 knots or 78 mph and service ceiling is 13,125 feet. Takeoff and landing run is 1640 feet and range with maximum payload is 459 nautical miles or 528 miles and with maximum fuel the range is 648 nautical miles or 745 miles.

Yak-130 is a two-seat light fighter/attack/reconnaissance combat airplane, advanced air defense and deck trainer, special mission and towing target drones It is an un-conventional all swept mid wing monoplane. Engines are located in ducts under the wing roots beneath LEREX extending almost to the windscreen. It has a design service life of 10,000 flying hours. The landing gear is retractable tricycle type with a single wheel on each unit. Main wheels retract into engine ducts and low-pressure tires are included. It is powered by two turbofan engines, each generating 4850 pounds st. Accommodations are for a crew of two seated in tandem under the blister canopy with the rear seat raised for better visibility. Seven hard points are provided for guns, missiles and guided or unguided bombs. Wingspan is 34 feet 11 inches and overall length is 39 feet ½ inch. Maximum internal fuel is 35527 pounds and with external tank, 4850 pounds. Normal takeoff weight is 13,230 pounds and maximum takeoff weight is 18,740 pounds. Maximum level speed at height is 512-540 knots or 590-620 mph, takeoff speed is 108 knots or 125 mph and landing speed is 105 knots or 122 mph. Service ceiling is 39,370 feet, takeoff run is 1250 feet and landing run is 2200 feet. Maximum ferry range with conformed tank is 1185 nautical miles r 1365 miles.

Sustained g limit at 15,000 feet altitude is +5, and at never exceed limits are +8 and –3.

PART III

Lighter-Than-Air Vehicles

The main Lighter than Air vehicles are commonly called Hot Air Balloons, Dirigibles, Airships or Blimps. The first flight of Hot Air Balloons carrying passengers was in 1783 in Paris, France. It was built with cloth bags lined with paper with a fire being built on a grill attached to the bottom. These were infrequently but had a tendency to be destroyed by fire upon landing. King Louis XVI originally declared that condemned criminals would be the first pilots, but a physicist named Pilatre de Rozier petitioned the King for the first honor. The previous flights were with unmanned with animals being the only occupants. The manufacturers, were Josef and Ethienne Montgolfier of Annonay France who were paper manufacturers and who had noticed the ash rising in fires. The first with humans on board took place in October 1783. In January, 1784 a huge hot Air Balloon carried seven passengers to a height of 3,000 feet over the city of Lyons.

Unmanned hot air balloons were used in China for military signaling. These were known as Kongming lanterns. It is also thought that the Nazka Indians in Peru used then about 1500 years ago. The first modern hot-air-balloon was designed and built in 1960 by Ed Yost. He made the first flight on November 22, 1960. It was initially equipped with a plastic envelope and kerosene fuel. These designs were rapidly improved to using a modified propane "weed burner" to heat the air and lightweight nylon fabric.

The first Military application of balloons was during the French Revolutionary War when the French used tethered balloons to observe the movements of the Austrian Army during the Battle of Fleurus in 1794. They were also used during the America Civil War. The military balloons were used by the Union Balloon Corps, commanded by Professor Thaddeus C. Lowe and were limp silk envelopes inflated with coke gas or hydrogen. The Confederate Army attempted to counter with a rigid Montgolfier style hot air balloon, commonly called the "hot smoke balloon." Captain John R. Bryant inflated his rigid cotton balloon with a fire of oil-soaked pine cones. The balloon was soon captured by the Union Forces as the Confederate's technique of balloon handling were not competent. Today, there are almost 8,000 hot air balloons used primarily for recreation in the United States. These hot air balloons are capable of high altitudes. In November, 2005, Vijaypat Singhania set a world record of 60, 852 feet or 21,290 meters. He started at Bombay, India and landed 150 miles south of Panchale. The previous record held was 64,980 feet by Per Lindstrom at Plano, Texas in June 1988. However, oxygen is needed for all passengers for any flight that reaches and exceeds 12,500 feet altitude as well as other Aircraft. On January 15, 1991 The Virgin Pacific Flyer completed the journey from Japan to Northern Canada, flown by Per Lindstrom and Richard Branson. The distance was 7,671.91 kilometers. With a volume of 2,600,000 ft cubed or 74,000 meters cubed this was the largest ever hot air balloon ever built. It was designed to fly in the trans-oceanic jet streams and recorded the highest ground speed for a manned balloon at 245 mph or 394 km/hr. The longest duration of a hot air balloon ever made was 50 hours and 38 minutes, flown by Michio Kanda and Hirosuke Tekezawa of Japan on January 2, 1997.

The current hot air balloon for manned flight uses a single layered fabric gas bag with an opening at bottom called the mouth or throat. A basket, commonly called the gondola is attached to the envelope for transport of passengers. The basket may be made of wicker and rattan, but is usually aluminum. Above the basket and cantered in the mouth is a burner which injects a flame into the envelope to heat the air. The burner is fueled by Propane stored in pressure vessels similar to high pressure forklift cylinders. Modern hot air balloons are made of fabrics such as rip stop nylon or Dacron. The material is cut into panels and sewn together along with structural load tapes that carry the weight of the gondola or basket. Vertical rows of panels referred to as gores due to their triangular shape. Envelopes can have as few as few as 4 gores or as many as 24 or more. More gores usually means a smoother shape. Envelopes have a crown ring at the top. this is a hoop of aluminum approximately 1 foot in diameter to which vertical load tapes are attached.

The fabric or at least the top one-third, is coated with a sealer such as silicone or polyurethane to make it impermeability to air. Heat, moisture and mechanical wear during set-up and packing are the primary causes of degradation. When an envelope becomes too porous to fly, it is usually retired and used for children to run through. Products for recoating the fabric are becoming available commercially.

Envelope sizes are range from one-person basketless balloons which have less than 35,000 cubic feet of envelope to balloons that carry well over 24 people used by large sightseeing operators and have envelope volumes up to 600,000 cubic feet. Most balloons are roughly 100,000 cubic feet and carry 3 to 4 people.

Burners are fired by LP [Liquid Propane] gas, when mixed with air, ignites the mixture and directs the flame and exhaust into the mouth of the envelope. The pilot may use one or more to generate the desired heat. Each burner is characterized by a metal coil of propane through which the flame shoots to preheat the incoming liquid propane. The burner may be suspended from the mouth or rigidly supported over the basket. The burner unit may be mounted on a gimbal ring to enable the pilot to aim the flame and avoid overheating the envelope fabric. The pilot actuates a burner by opening a propane valve called a "Blast Valve". This valve may be spring-loaded so it can close automatically or the pilot can make the valve stay open until he wants it to close. The burner has a pilot light to ignite the air and propane mixture. The burner may have a secondary valve that releases propane more slowly and generates a different sound, called a "Whisper Burner." It is used for flight over livestock to lessen the noise which may "spook" them.. It also generates a yellow flame and is used for night glows because it lights up the inside of the envelope better than the primary valve. Burners can generate heat of about 30 million BTU. Hot Air Balloons are made in many different shapes and sizes. The average system weight is as follows:

Envelope, 250 pounds
Basket, 140 pounds
Burner, 50 pounds
3 fuel tanks with fuel, 405 pounds
5 passengers, 750 pounds
Air in the envelope [100,000 cubic feet @ 0.062 pounds per cubic feet] 3.1 tons
TOTAL WEIGHT = 3.9 tons

The largest manufacturer of hot Air balloons is Cameron Balloons of Bristol, and a subsidiary company Lindstrand Balloons of Oswestry, England. North Americas largest was Aerostar International of Sioux Falls, South Dakota but they quit making balloons in January Firefly Balloons, formerly known as The Balloon Works, another popular manufacturer, is located in Statesville, North Carolina. Another long time producer of hot air balloons is Head Balloons, Inc of Helen, Georgia.

AIRSHIPS/DIRIGIBLES

Early balloons were not navigable. Trials to make them improve their maneuverability included elongating the balloon's shape and using a powered screw to push it through the air. Thus, the airship, also called a dirigible was born. In 1852, a French engineer, Henri Giffard attached a small steam engine to a huge propeller and flew for seventeen minutes at a top speed of 5 miles per hour. After the gasoline engine was invented in 1896 airships could be built. In 1898 the Brazilian, Alberto Santos-Dumont, constructed and was the first to construct and fly a gasoline powered airship. Arriving in Paris in 1897, he first made a number of flights with free balloons and also purchased a motorized tricycle. His plan was to combine the De Dion engine with a balloon which resulted in 14 airships, all gasoline powered. His number 1 airship first flew in September 1898.

Zeppelin was the name given to the duralumin-internal-framed dirigibles invented by Count Ferdinand Zeppelin. The first rigid airship flew on November 3, 1897 and was designed by David Schwarz who was a timber merchant. Its skeleton and outer surface was made of aluminum and powered by a 12 horsepower Daimler gasoline engine connected to three propellers. It lifted successfully upward in a tethered test near Berlin, Germany, but crashed. Ferdinand Zeppelin was born in 1838 and died in 1917. He invented a rigid frame dirigible that became known as the Zeppelin. He flew the first un-tethered airship named the LZ-1 on July 2, 1900 near Lake Constance in Germany. The cloth-covered dirigible was the prototype of many subsequent models. It had an aluminum structure, 17 hydrogen cells and two Daimler internal engines, each driving two propellers. It was about 420 feet long and 38 feet in diameter. In its first flight, it flew 3.7 miles in 17 minutes and gained altitude to 1300 feet. In 1908 Mr. Zeppelin established the Zeppelin Foundation for development of aerial navigation and the manufacture of airships.

In the 1930s Goodyear built two giant airships for the U.S. Navy. They had internal metal frames to maintain their shape. The airships measured more than two football fields and needed 6.5 million cubic feet of helium and to become airborne and the gross weight was more than 400,000 pounds. Both airships were lost within two years, ending the era for rigid airships. In the 1940s and 1950s Goodyear built a series of surveillance airships to protect merchant fleets along the coast of the U.S. They were also used for early warning radar stations. They could stay aloft for more than a week at a time. An airship of this kind still holds the flight endurance record of 11 days in flight. It was a Goodyear airship called the ZPG-2 Snow Bird and was flown from Weymouth, Massachusetts to Europe and Africa and return to to Key West, Florida without refueling or landing.

The U.S. Navy had two ZP squadrons at Weeksville Naval Air Facility, North Carolina and it was there that the largest wood structure in the world was located. The Navy discontinued operation of airships in 1962.

Meanwhile, Europeans were for the first time ever could fly by dirigible to the U.S. instead of traveling by passenger ships. The Hindenburg was built by Zeppelin. On May 6, 1937. Hindenburg, a luxurious flying hotel and faster than any ship, 800 feet long and carrying 36 passengers crashed at Lakehurst, New Jersey, killing 35 of 36 on board. There are still questions as to why it crashed, but no one is sure. It was assumed at that time that lightening ignited the hydrogen and caused the fire. Some thought that this blimp was sabotaged by people who opposed the Nazi regime that had recently gained power. More recently, Engineers and Scientists have theorized that a buildup of static electricity generated the spark that caused the explosion. The fact that the blimp remained largely intact suggests that it was the outer coating was extremely flammable and was struck by lightening and ignited the coating instead of the Hydrogen gas that filled the airship. It still remains undeterminable exactly what caused the explosion as all the evidence was destroyed.

Whatever caused that disaster, Zeppelin ceased operation in 1940 and most of their passengers and crews have diminished, but what they did and experienced lives on. Today, Zeppelin has returned, and built a new airship in 1997, the LZNT. It has been certified and passenger flights began in august 2001.

Today, the Goodyear Tire and Rubber Company no longer mass produces airships. In the United States, they operate three blimps, one based in Carson, California, one in Akron, Ohio, and the other in Pompano Beach, Florida. Over the years, Goodyear built more than 300 airships which is more than any other company in the world. Akron, Ohio was the center of manufacturing for several decades.

Over the years many have speculated where the term Blimp came from. As the story goes, Lt. D. Cunningham of Great Britain

Air Service, commanded the air station at Capel, England during world War I. While conducting a weekly inspection of the station, he flipped his thumb at the envelope of His Majesty's Airship, SS-12, and an odd noise echoed from the taut fabric. He cried, "Blimp" imitating the sound sand the rest is history.

APPENDIX A

Definitions

AAH-Automatic Approach to Hover.

AC-Alternating Current.

ACE-Actuator Control Electronics.

ACLS-Automatic Carrier Landing System: Air Cushion Landing System.

ACMI- Air Combat Maneuvering Instrumentation.

ADF-Medium Frequency Automatic Direction Finding.

ADG-Accessory Drive Generator.

ADI-Attitude/Director Indicator.

AEW-Airborne Early Warning.

AFCS-Automatic Flight Control System.

AGM-Air To Ground.

AH-Ampere-Hours.

AHRS-Altitude/Heading Reference System.

AIDS-Airborne Integrated Data Systems.

AIS-Advanced Instrumentation System.

ACLM-Air Launched Cruise Missile.

ANHEDRAL-Down slope of wing seen from front in direction from root to tip.

ANVIS-Aviator's Night Vision System.

AOA-Angle Of Attack.

APU-Auxiliary Power Unit.

ASE-Automatic Stabilization Equipment OR Aircraft Survivability Equipment.

ASI-Airspeed Indicator

ASM-Air-To-Surface.

ASPJ-Advanced Self-Protection Jammer.

ASW-Anti-Submarine Warfare.

ATDA-Airborne Tactical Data System.

AUW-All Up Weight [Total weight of aircraft under defined conditions, or at a specific time during flight]. Not to be confused with MTOGW.

AWACS- Airborne Warning and Control System.

BITE-Built-In Test Equipment.

CHAFF-Thin slivers of radar reflective material cut to appropriate lengths of wavelengths of hostile radars and scattered in clouds to protect friendly aircraft.

COMINT- Communications Intelligence.

COMPOSITE MATERIALS-Made of two constituents, such as filaments of short whiskers plus adhesive forming a binding matrix.

C/R- Counter rotating propellers.

CRT-Cathode Ray Tube.

DADS-Digital Air Data System.

Db-Decibel.

DIHEDRAL-Upward slope of wing seen from front, in direction from root to tip.

DINS-Digital Inertial Navigation System.

DME-Distance Measuring Equipment.

DRONE-Non-piloted vehicle aircraft.

ECS-Environmental Control System.

ELEVON- Wing trailing-edge control surface combining functions of aileron and elevator.

ESM-Electronic Surveillance Measurement/Electronic Signal Monitoring.

EW-Electronic Warfare.

FBW-Fly By Wire [Cables and mechanical controls are replaced by electrical wiring.

FEATHERING-Setting propeller or similar blades at pitch aligned with slipstream, minimizing drag.

MFD-Multi-function [electronic] display.

FEBA-Forward Edge of the Battle area.

FMS-Foreign Military Sales OR Flight Management System.

RADOME- A large bubble to house radar.

FERRY RANGE-Extreme Safe Range with zero payload.

FIELD LENGTH-Measure of distance needed to land and/or takeoff.

FLAT FOUR/FLAT SIX-Engine having four or six Horizontally opposed cylinders.

FLIR-Forward Looking Infrared.

FLY BY LIGHT-Flight control system in which signals pass between computers and actuators along fiber-optic leads.

FLY BY WIRE-Flight control system with electrical leads and without mechanical connection.

FM-Frequency Modulation.

FREE TURBINE-Turbine independent of engine upstream, connected by rotating bearings and the gas system and able to run at its own speed.

FSW-Forward swept Wing.

FY-Fiscal Year.

g-Acceleration due to earth gravity, of a body in free fall or due to a rapid change of direction of Flight Path.

GLOVE-Fixed portion of wing inboard of variable swept wing, or additional airfoil profile added around normal wing for test purposes.

GPS-Global Positioning System satellite –based precision navigation aid.

GSE-Ground Support Equipment.

GUNSHIP-Helicopter designed for battlefield attack, with a slim body and carrying a pilot and gunner only.

HF-High Frequency.

HMD-Helmet mounted display.

HOVER CEILING-Altitude ceiling of a Helicopter.

HP-Horse Power.

HUD-Heads Up Display.

HZ- Cycles per Second.

IAS-Indicated Air Speed.

IFF-Identification, Friend or Foe.

IFR-Instrument Flight Rules.

INS-Inertial Navigation System.

ILS-Instrument Landing System.

IMC-Instrument meteorological Conditions.

ISA-International Standard Atmosphere.

IMPERIAL GALLON-1.20095 U.S. Gallons or 4.546 liters.

INS-Inertial Navigation System.

INTEGRAL TANK-Liquid tank formed by sealing part of the structure.

INVERTER- Electric or electronic device for reversing polarity alternate waves in AC to DC.

IR-Infrared.

JATO-Jet Assisted Takeoff.

KEVLAR-Aramid fiber used as high strength composite material.

KNOT-one nautical mile per hour

LCD-Liquid Crystal Display.

LED-Light-Emitting Diode.

MACH NUMBER-The ratio of the speed of a body to the speed of sound.

MAD-Magnetic anomaly detector.

MLS-Microwave Landing System.

MMS-Mast Mounted Sight.

MRW- Maximum Ramp Weight.

MTOGW- Maximum Takeoff Gross weight. [MRW minus taxi/run-up fuel].

MONOQUE-Structure with strength in the outer shell, devoid if internal bracing.

MPH-Miles Per Hour.

NASA-National Aeronautics and Space Administration.

NATO-North Atlantic Treaty Organization.

NDB-Non Directional Beacon.

NM-Nautical Mile.

NOE-Nap of the Earth, Low level flying using the natural cover of hills, trees, etc. and following the contour of the earth.

NVG-Nigh Vision Goggles.

OEI-One Engine Inoperative.

OGE- Out of Ground Effect, Helicopter hovering far above the nearest surface.

PAYLOAD-Disposable load generating revenue in military/civil aircraft, or total load carried of weapons cargo, etc.

PLUME-The region of hot air and gas emitted by a jetpipe.

PORT-Left side looking forward..

PROPFAN-A family of new technology propellers characterized by multiple scimitar-shaped blades with thin sharp edged profile.

RAMP WIGHT-Maximum weight at the start of flight.

RANGE-The distance an aircraft can fly with a specified load and making allowance for additional maneuvers.

RPM-Revolutions Per Minute.

S-Seconds.

SAR-Search and Rescue.

SERVICE CEILING-Height equivalent to air density at which maximum rate of climb is 100 feet per minute.

SHP-Shaft Horsepower, measure of power transmitted via a rotating shaft.

SIGINT-Signal Intelligence.

S/L-Sea Level.

SLAR-Side looking airborne radar.

SST-Supersonic Transport. st-Short Tons, State, stet, Stokes (Unit of weight)

STALLING SPEED- Airspeed at which aircraft stalls at eg, wing lift suddenly collapses.

STARBOARD-Right side, looking forward.

STOL-Short Takeoff and Landing.

STROBE LIGHT-High intensity flashing beacon.

SWEEPBACK-Backward inclination of the wings from above.

TABS-Small auxiliary surfaces hinged to trailing edge of control surfaces of trimming.

TACAN-Tactical Air Navigation UHF military navigation aid giving bearing and distance element.

TILT ROTOR-Aircraft with fixed wing and rotors that tilt up for hovering and forward for fast flight.

TURBOFAN-Gas Turbine jet engine generating thrust by a large diameter cowled fan, with a small part added by jet by from core.

TURBOJET-Simplest form of gas turbine comprising compressor, combustion chamber, turbine and propulsive nozzle.

TURBOPROP-Gas turbine in which as much energy as possible is taken from gas and used to drive a reduction gearbox and propeller.

TURBOSHAFT-Gas turbine in which as much energy is taken

from gas jet and used to drive a high-speed shaft which in turn drives an external load such as a helicopter transmission.

UAV-Unmanned air vehicle.

UHF-Ultra High Frequency.

USABLE FUEL-Total mass of fuel consumable in flight, usually 95-98 prercent of system capacity.

USEFUL LOAD-Usable fuel plus Payload.

VARIABLE GEOMETRY-Capable of grossly changing shape in flight, especially by varying the sweep of wings.

VFR-Visual Flight Rules.

VHF-Very High Frequency.

VTOL-Vertical takeoff and Landing.

ZERO/ZERO seat-Ejection seat designed for use even at zero speed on the ground.

CPSIA information can be obtained at www.ICGtesting.com
Printed in the USA
238720LV00002B/27/P

9 781612 040936